TODAY'S ARCHITECTURAL MIRROR:
Interiors, Buildings, and Solar Designs

TODAY'S ARCHITECTURAL MIRROR:
Interiors, Buildings, and Solar Designs

Pamela Heyne

VAN NOSTRAND REINHOLD COMPANY
NEW YORK   CINCINNATI   TORONTO   LONDON   MELBOURNE

Van Nostrand Reinhold Company Regional Offices:
New York   Cincinnati   Atlanta   Dallas   San Francisco

Van Nostrand Reinhold Company International Offices:
London   Toronto   Melbourne

Copyright ©1982 by Van Nostrand Reinhold Company

Library of Congress Catalog Card Number: 81-10375
ISBN: 0-442-23424-4

Manufactured in the United States of America

Published by Van Nostrand Reinhold Company
135 West 50th Street, New York, N.Y. 10020

Published simultaneously in Canada by Van Nostrand Reinhold Ltd.

15 14 13 12 11 10 9 8 7 6 5 4 3 2 1

Library of Congress Cataloging in Publication Data

Heyne, Pamela.
  Today's architectural mirror.

  Bibliography: p.
  Includes index.
  1. Mirrors in architecture. I. Title.
NA2796.H49   729   81-10375
ISBN 0-442-23424-4   AACR2

To My Parents
and Peter

Theme float at the head of Jimmy Carter's presidential inaugural parade, 1977. In reflecting the images of the observers, the mirrors were to emphasize the importance of the people themselves in the American form of government. The float itself bore a resemblance to mirror glass buildings. (*Courtesy The White House*)

Introduction

Not too long ago, the mirror was an unpopular architectural material. A generation of designers in the middle of this century largely ignored it. Too ambiguous or illusionary or flashy, it did not fit easily into the philosophical or built environment of the time. Today, however, it is ubiquitous. We see it, not just in bathrooms and entrance foyers, but banks, offices, kitchens, and even a church. In the form of mirror glass, a semitransparent mirror, it clads entire office and hotel complexes. And here and there, we see architects experiment with the technical solar mirror. Suddenly, the mirror has become not just an interior architectural material, but a highly visible, exterior material as well.

As far as the history of the architectural mirror is concerned, the period of the 1950's is akin to a watershed. Prior to that time, architects used the mirror on interiors. After that time, they could sheath multistory buildings with it. And the solar mirror could provide much of the building's energy requirements.

So today we see more architects using the mirror, in a greater diversity of ways, than at any time in the past. This book's purpose is to answer two questions, why and how. Why has the mirror become so widespread, particularly in the United States? And exactly how are designers using it?

The first half of the book deals with the interior mirror.

Today, it satisfies frivolous needs as well as serious ones. It might add sparkle to a discotheque, or illusionary space to a too-cramped apartment. And while architects are experimenting with this material, they are often duplicating techniques first developed centuries ago. Today's designers undoubtedly consider their effects highly original, yet they might be based on techniques first used in European palaces, or even early magic shows.

The second half of the book investigates the mirror glass building and the solar mirror. The basic *raison d'être* of these mirrors is to reflect heat-producing sunlight. Mirror glass, sometimes called reflective glass, was developed to keep building interiors cooler by reflecting sunlight away before it even enters the building. The solar mirror focuses solar energy, creating various levels of usable heat. It can warm and cool buildings, power industrial processes, and create electricity.

The emphasis of this phase of the book is the mirror glass building. It is so widespread and so visible, and still a new form, architecturally speaking. From an aesthetic standpoint, it is an extremely difficult form to execute well. A mirror covering all four sides of a 50-story building means a lot of a surface that one typically will look "at" rather than "through," a surface that is often bright and glaring, a surface that all too often reflects cars and freeways instead of simply cumulus clouds. For these reasons, the mirror glass building has not been enthusiastically received by most critics. Arthur Drexler termed it abstract and remote. Peter Blake wrote he hoped our future cities wouldn't be filled with these shiny facades. Rod MacLeish equated it with slick American commercialism, saying, "Now comes word from Hollywood that two film companies are remaking versions of King Kong. That is outrageous. It is equivalent to remodeling the Parthenon to include a McDonald's arch and one-way mirror windows."

Techniques for dealing with so much mirror have emerged. Some have derived from the interior mirror. Most techniques however are new, recently created by architects as a way of taming this chameleonlike material. The few mirrored buildings that have succeeded aesthetically have done so because their designers realized the peculiar limitations and problems with a mirror out-of-doors. They have understood the need for a certain amount of restraint in designing with such large mirrors.

The interior mirror, the mirror glass building, and the solar mirror are all basically the same material. They are just used differently, for different purposes, and hence seem to be different things. But they are all unified by their ability to reflect light more perfectly than any other material. As that reflected ray emerges in a predictable fashion, the designer can control it. It can be shifted, multiplied and focused at will.

In the middle of this century, few designers took the mirror seriously. Now, many architects realize that an illusionary material is not necessarily a frivolous material. Nor is it simply an interior material.

Acknowledgments

I wish to express my thanks to those who provided illustrative material. I am also appreciative of extensive interviews granted me by Peter Blake, Kevin Roche, John Dinkeloo, Cesar Pelli, Robert A.M. Stern, Charles Gwathmey, Hugh Jacobsen, Henry Cobb, Paul McGowan, and many others. And finally, thanks go to Mary Frances Diedrich for assistance, and Carole Gottschalk, who simply sent a timely postcard that said "How's the book?"

Contents

List of Color Plates

PLATE 1. Richardson-Merrell, Inc. Headquarters, Wilton, Connecticut, 1974.

PLATE 2. John Deere Co., Moline, Illinois, 1978.

PLATE 3. United Nations Plaza restaurant, New York, 1975.

PLATE 4. Shezan restaurant, New York, 1976.

PLATE 5. Shezan restaurant, New York, 1976.

PLATE 6. Providence Journal Headquarters, Providence, Rhode Island, 1976.

PLATE 7. Gund Investment Co. Office, Princeton, New Jersey, 1980.

PLATE 8. Greenwich Savings Bank branch, New York, 1975.

PLATE 9. Residence, Westchester County, New York, 1976.

PLATE 10. Infinity Chambers, New York, 1978.

PLATE 11. Xavier's apartment, New York, 1978.

PLATE 12. Bell Labs, Holmdel, New Jersey, 1962–1967.

PLATE 13. John Hancock Tower, Boston, Massachusetts, 1976.

PLATE 14. Canadian Pavillion, Expo '70, Osaka, Japan, 1970.

PLATE 15. California Jewelry Mart, Los Angeles, 1968.

PLATES 16, 17 and 18. Reunion Center, Dallas, Texas, 1978.

PLATES 19 and 20. Hampden Country Club, Springfield, Massachusetts, 1974.

PLATE 21. Indiana Bell Telephone Switching Center, Columbus, Indiana, 1978.

PLATE 22. Indiana Bell Telephone Switching Center, Columbus, Indiana, 1978.

PLATE 23. United States Embassy, Tokyo, 1972.

PLATE 24. Citicorp Center, New York, 1979.

PLATE 25. Johns-Manville World Headquarters Building, near Denver, 1976.

PLATE 26. Johns-Manville World Headquarters Building, near Denver, 1976.

PLATE 27. Solar furnace at Odeillo, France.

PLATES 28 and 29. Vacation House, Carlyle, Illinois, 1978.

PLATE 30. Yocum Lodge, Snowmass at Aspen, Colorado, 1973.

PLATE 31. Strawberry Fields Apartments, Springfield, Missouri, 1976.

TODAY'S ARCHITECTURAL MIRROR:
Interiors, Buildings, and Solar Designs

1
The Interior Mirror

Oh, Kitty, how nice it would be if we could only get through into the Looking Glass House! I'm sure it's got, oh! such beautiful things in it! Let's pretend there's a way of getting through into it, somehow, Kitty. Let's pretend the glass has got all soft like gauze, so that we can get through.

Lewis Carroll, *Through the Looking Glass*

A mirror is only a shiny surface. It is usually a piece of very flat glass with a metallic backing. In most cases, the backing is silver. Perhaps the mirror is a total depth of 1/8". Yet within that plane, the illusion of infinity can be contained.

The essence of the mirror is its bright silver backing; the main function of the glass is to stabilize the thin silver film, and to protect it. Light will be reflected from this shiny smooth surface according to the law of reflection which states that the angle of incidence of any light ray is equal to its angle of reflection. Because of this law, we see images "in" a mirror (Fig. 1-1). Light which emanates from an object will strike the mirror, then bounce off its surface at an equal angle but opposite direction, and then into our eyes. Since the light emerging from the mirror contains the same messages, only reversed, as the light striking the mirror, we see "mirror images."

A "one-way" mirror differs from a regular mirror essentially in that it has a thinner reflective backing. Not all of the light striking the glass is reflected. Some is transmitted, allowing visibility through the glass under certain lighting conditions (Fig. 1-2).

Other materials have mirrorlike qualities. Glass itself will reflect some of the light that strikes it in a regular

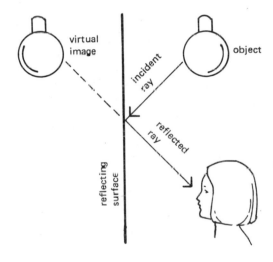

Fig. 1-1. The law of reflection. The angle of the reflected ray of light is equal to the angle of the incident ray.

1

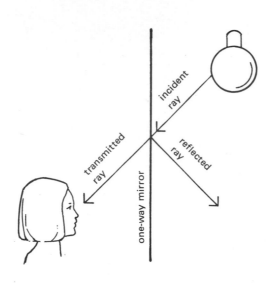

Fig. 1-2. A one-way mirror allows some light to penetrate the glass while some is reflected according to the law of reflection.

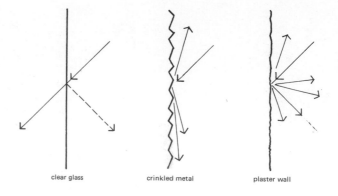

clear glass   crinkled metal   plaster wall

Fig. 1-3. Reflective characteristics of different materials.

fashion according to the law of reflection. But since most of the light that strikes it is transmitted, the reflections are very dim. A crinkled metal surface will form tiny images, according to the law of reflection, on each bent surface. The overall effect is one of distorted images. Dull, matte surfaces will not create images since the light is reflected in a diffuse fashion (Fig. 1-3).

The images that form on the surface of the mirror are so realistic that we forget about the presence of the surface itself on certain occasions. We have all had the experience of talking to a person while actually looking at the person's image in the mirror. The person's voice is obviously not coming from the surface of the mirror. Yet, the image is so compelling we continue talking to it.

Superstitions developed over the centuries based on the realistic appearance of images in the mirror. In the past, one's reflection in the mirror was often considered his soul. Nervous indeed was the literary figure who did not see his reflection in the mirror. In certain cultures, mirrors in the room of a sick person were often covered with a dark cloth so that the soul of the weakened person would not depart from the mirror. And even today, if we break a mirror, our knee-jerk reaction is to blurt out, "Seven years bad luck!"

Of course, when we look at a mirror, we see the reverse of what is actually in front of it. Each of us has a mental image of our own face which is the opposite of what it really is, thanks to the mirror. Generally, this reversal does not disturb us unless we see writing reflected in the mirror.

An object that is reflected in the mirror will seem as far back from the surface of the mirror as it is in front of the mirror. For instance, a vase two feet in front of the mirror will seem two feet back in the mirror's illusionary space. Interestingly, if we focus a camera on a reflection in a mirror, it will have to be set for the apparent distance of the reflected

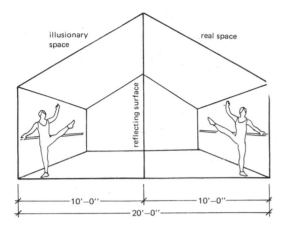

Fig. 1-4. A mirror can seemingly double a space, and architectural elements as well.

object, rather than the real distance of the mirror's surface.

In architectural terms, this apparent distance of objects is significant. If a wall is 10 feet from the reflecting surface, another wall will appear in the mirror's illusionary space, seemingly 10 feet behind that plane. The total space then will appear 20 feet wide, real space plus illusionary space. And architectural forms, such as a semicircular ceiling, can be completed in the mirror (Fig. 1-4).

Sometimes the illusions playing on the mirror's surface are less important in a particular design than bright shafts of light that emerge from its surface in a predictable fashion. A dark corner can be brightened with strategically placed mirrors (Fig. 1-5).

These two extremes, an illusionary mirror and a bright surface, represent two distinct ways of thinking about the mirror. On one hand, it can be so illusionary that we forget totally we are looking at a mirror. On the other hand, it can be a visible surface, a surface that creates illusions, but with a palpable existence of its own. These two distinct attitudes can encompass a whole host of techniques.

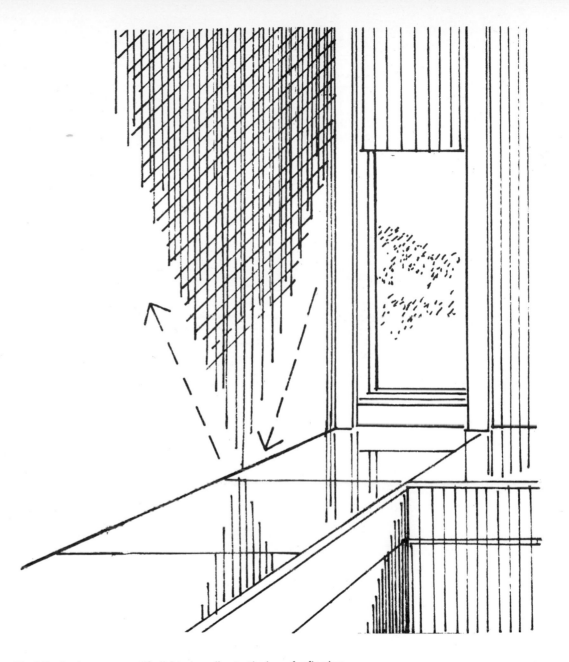

Fig. 1-5. A mirror can amplify light, according to the law of reflection.

3

BACKGROUND

The mirror made of glass with a reflective backing is very much a man-made item. In nature, the closest approximation to the mirror is the still pool of water. The myth of Narcissus is about a youth who fell in love with his reflection in a river. Punished by the gods for his vanity, his body was eventually transformed into the flower that bears his name, the narcissus or daffodil.

The first man-made mirrors were actually polished stone, such as obsidian or phengite, a volcanic glass with a high sheen. Such mirrors have been found in Chinese and Incan ruins. As man developed the skill to extract metals from rock, he learned to impart a mirrorlike sheen to the metals. In the ancient civilizations of China, Egypt, Greece and Rome, bronze became the most common form of mirror. Around the Mediterranean, gold and silver were occasionally fashioned into mirrors (Fig. 1-6).

Such early mirrors were generally small, used as an aid in personal adornment. A typical example would be approximately 6″ in diameter with an elaborately decorated back and handle. They had no usefulness as wall mirrors in a decorative sense. Rather, they were similar to our hand mirrors of today.

As long as mirrors remained made of metal exclusively, they were destined to remain small in size, for metal was expensive and heavy. If metal were fashioned into thin large sheets, it would warp and the reflections would be distorted. Not until metals could be made to adhere to clear, high quality glass could mirrors of any size occur.

Glass mirrors were known in the ancient world but were rarely made. Pliny, the Roman naturalist, spoke about glass which was coated with tin or silver at Sidon. Given the irregular quality of ancient glass, however, such mirrors did not have the requisite smoothness and brightness to form perfect images.

Glass, a compound silicate of soda and sand, or soda, sand and lime (the lime made it waterproof), was an imperfect product in ancient times. The glass for windows

Fig. 1-6. Chinese bronze mirror, back decorated with gold repoussé bird-and-flower pattern. T'ang dynasty, 618-907 A.D. Diameter 8-11/16 inches. Centuries would elapse before the front of the mirror was considered a decorative element. (*Courtesy the Freer Gallery of Art, Washington, D.C.*)

or mirrors was typically made by a crude method of casting on the ground. It had a dark color and was irregular. Centuries would elapse before glass of consistent quality and hence, mirrors, could be produced.

Even in the Middle Ages, though glass mirrors were known, the most common mirrors, called specula, were made of steel or silver. They often dangled from a woman's waist, as an aid to grooming.

The first decorative wall mirrors were produced at Nuremberg during the twelfth and thirteenth centuries AD. The craftsmen, highly skilled in the art of blown glass, would introduce molten lead or tin into the still hot blown forms. The glass globes were subsequently rotated until a thin film of metal would completely cover them. They were then cut to size when cool. They were known as the convex Nuremberg mirrors. A well-known example exists

Fig. 1-7. Titian (Venetian, c. 1477–1576): *Venus with a Mirror,* painted c. 1555. While Venus is contemplating her face in the mirror, it is not a small hand mirror but one that could be attached to a wall. The cupid is clearly holding it with some difficulty. (*Andrew W. Mellon Collection. Courtesy National Gallery of Art, Washington, D.C.*)

in the painting by VanEyck of Giovanni Arnolfini.

Nuremberg's dominance gave way to Venice in the sixteenth and seventeenth centuries. As with Nuremberg, Venetian glass began in a blown form. But the Venetians would blow the glass into a long, cylindrical shape. It would then be slit and opened out to create a relatively flat sheet. It would then be annealed, a process of heating and gradual cooling. After that, it had to be ground and polished on both sides by hand.

The final process in turning glass into mirror was known as "foiling." It consisted of making a sheet of tin foil adhere to the back of the glass. The foil was spread out and coated with an amalgam of mercury. The glass was then placed on top of the amalgam, weighted, tilted and left to drain for nearly two weeks.

This process was obviously laborious and time-consuming. It was also a great secret for some time. The Venetians prevented other nations from obtaining their knowledge through strict measures. The glassworks were located on the island of Murano. Venetian craftsmen were forbidden to travel, to reveal their trade secrets, even to export glassmaking tools or bits of broken glass.

A Venetian wall mirror was often beveled and elaborately framed. Sometimes the frame itself was made of mirror. The size was typically what we might find over a dressing table today. The mirror, after thousands of years as being a small accessory to personal adornment, was becoming an architectural tool (Fig. 1-7).

The first large scale use of the mirror occurred in France, although the mirrors were imported from Venice. The Hall of Mirrors at Versailles, designed for Louis XIV by Charles Le Brun in 1682, was a lavish concept indeed. It consisted of monumentally arched openings, filled with mirrors, facing large windows. Over 300 mirrors of approximately 2½' x 3½' were used in the scheme. It is little wonder that Louis XIV often destroyed his bills for Versailles so that his ministers would not discover exactly how much it cost (Fig. 1-8).

Very quickly, high quality glass and mirrors were in great demand by the nobility. They wanted large sheets of glass for their coaches and mirrors for their salons and places of entertainment. At the end of the seventeenth century, the French ended the Venetian monopoly in high-quality glass products through improved manufacturing techniques.

In 1687, the glassmaker Bernard Perrot invented a process for making plate glass by casting. The process was further developed by Louis Lucas de Nehou in the following year. Glass would emerge from the oven in molten form and be poured directly onto a large casting table. After the glass cooled, it still required grinding and polishing by hand, but to a lesser extent than blown glass. As a result, sheets of glass as large as 4' x 7' could be made. The de Nehou process of casting as it·came to be known lasted well into the twentieth century.

Fig. 1-8. Charles Le Brun: Hall of Mirrors at Versailles, France, 17th century. (*Courtesy French Government Tourist Office*)

Voltaire, writing in the *Age of Louis XIV*, noted the improvements in French mirror manufacture. He said, "... Mirrors as fine as those of Venice, which had always supplied the whole of Europe, began to be made in France; they attained a size and beauty which has never been successfully imitated elsewhere."

Today, mirrors are so commonplace and easy to buy that it is difficult to comprehend the esteem with which they were held in the seventeenth and eighteenth centuries. Yet, one does catch a glimpse of their value in another passage by Voltaire in his history. He considered mirrors to be among the most significant developments of the Renaissance. He described the years which preceded the reign of Louis XIV as being devoid of virtue:

So for almost nine hundred years, the genius of the French was almost continuously inhibited by a Gothic government, plagued by divisions and civil wars, lacking fixed laws and customs, changing, every two hundred years, its perpetually untutored tongue. Its nobility were undisciplined, knowing only war and idleness; its churchmen lived in disorder and ignorance; its peoples, bereft of industry, stagnated in poverty.

The French had no part in the great discoveries and the admirable inventions of other nations. Printing, gunpowder, mirrors, telescopes, proportional compasses, the air pump, the true system of the universe—all these were no property of theirs.

Mirrors, and for that matter, the air pump, were placed on the same level as the true system of the universe. By the eighteenth century, they could be found in abundant use in courts throughout Europe.

A vocabulary of mirror design began to develop. As at Versailles, mirrors were often placed opposite windows in order to amplify the existing daylight. They were often

DECORATION D'UNE SALLE D'ASSEMBLÉE VÜE DU CÔTÉ DES CROISÉES

A . Croisée a banquette et a double parement dont les developpemen se voyent
  a la planche 97.
B . Venteaux de parte Croisée agrandis panneaux pour recevoir des glaces.
C . Partie d'un des venteaux de la Croisée a banquette.
D . Guichet ferré sur le dormant de la Croisée, et tous fermé
  par l'espagnolette.

Echelle de Six Pieds.

E . Parement des guichet brisé.
F . Espagnolette fermée sur la Croisée et qui paroisse le guichet D sur le Dormant
  de la Croisée.
G . Banquette de menuiserie revêtue de marbre.
H . Embrasemens contre lesquels se viennent range les guichets lorsquils
  contreverte et brisée.

Fig. 1-9. Pier mirrors, French, 18th century. These mirrors lightened visually the wall areas between the windows. From J.F. Blondel, *De la Distribution des Maisons de Plaisance.*
(*Courtesy Library of Congress*)

7

placed between windows in the form of pier mirrors. Pier mirrors made the massive outer walls seem less heavy. In architecturally awkward spaces, they were particularly useful. If, for instance, there were only two windows along a wall with an intervening pier, a mirror on the pier would help to dematerialize it and make it less noticeable (Fig. 1-9).

Another important location for the mirror was over fireplaces. Called the overmantel mirror, it was useful in imparting a brightness to a portion of the room in which there were invariably no windows. It also helped to diminish the mass of the fireplace. In 1737, the designer J.F. Blondel mentioned in some of his writings that the fireplace was the most important area in a room on which to concentrate decoration, particularly if one did not have a great deal of money to spend. On the other hand, he noted that if one wished to get an idea of how much wealth another person might have, one merely had to observe how elaborately his fireplace was decorated. In the eighteenth century, the fireplace was a social indicator of sorts. When owners could afford it, it was invariably mirrored (Fig. 1-10).

As the quality of mirrors improved and the size of the panes increased, they were often placed opposite other mirrors. This created the effect called *"glaces à répétitions"* or vista mirrors. It was important for the glass to be quite smooth, otherwise distortions would build up very quickly in the reflected images. Invariably, there was a chandelier placed between the vista mirrors so that when it was lighted, there was an expansion of light into infinity rather than darkness. The chandelier compensated for the naturally greenish tint of the glass in mirrors to be intensified when reflected *ad infinitum.*

Writing in the twentieth century, Edith Wharton described in *The Decoration of Houses* the grandeur of the vista mirror:

Fig. 1-10. An overmantel mirror, French, 18th century. These mirrors were frequently placed face-to-face, resulting in *"glaces à répétitions"*. (*From J.F. Blondel, Courtesy Library of Congress*)

The old French decorators relied upon the reflection of mirrors for producing an effect of distance in the treatment of gala rooms. Above the mantel, there was always a mirror with another of the same shape and size directly opposite; and the glittering perspective thus produced gave to the scene an air of fantastic unreality. The gala suite being so planned that all the rooms adjoined each other, the effect of distance was further enhanced by placing the openings in line, so that on entering the suite it was possible to look down its whole length. The importance of preserving this long vista, or enfilade, as the French call it, is dwelt on by all old writers on house decoration.

The Amalienburg Pavilion by François Cuvillies is a memorable example of *glaces à répétitions*. Since the mirrors are deployed in an arc, we get oblique views of windows and openings, as well as other mirrors. Another important aspect of the Amalienburg is the gilt decoration. Though the Flemish-born Cuvillies had studied rococo design in Paris, this work is far more exuberant than any the French designed. The decoration floats over column and entablature, breaking it up almost as much as the mirrors do (Fig. 1-11).

With the eventual decline of the aristocracy at the end of the eighteenth century, mirrored installations approaching the lavishness of the Amalienburg were more likely to be found in public settings. The new restaurants and opera houses had their mirrors and gilt frames and decorated ceilings.

The restaurant first began in France after the revolution. It was more elaborate than the earlier inns and taverns, with a more refined menu. It permitted the new bourgeoisie to dine as well as the aristocracy had, if only for the duration and price of one meal. Multicourse banquets were concocted by chefs such as Carême, and served in settings filled with mirrors and gilt frames. Sometimes banquettes were placed along one or more walls and backed with mirrors, so that all diners could gain an improved view. As the restaurant

Fig. 1-11. François Cuvillies: Amalienburg Pavilion, Nymphenburg Park, Munich, 18th century. A lavish example of *glaces à répétitions*. Photo Sandak.

9

Fig. 1-12. Music Room at the Palace of Fontainebleu, French, 19th century. With improved size and quality, mirrors often were highly illusionary. (*Courtesy French Government Tourist Office*)

spread to other countries, so did the "French style" of banquettes and mirrors.

Mirrors became more visible and available to the general public in the nineteenth century. One important factor was a continual improvement in manufacturing techniques. In the 1840's, a German chemist, Justus von Liebig, developed a revolutionary method for applying the reflective coating to the back of the glass. Eliminating the cumbersome foiling process, Liebig was able to pour a liquid solution of silver directly to the glass. It improved reflectivity and decreased production time.

Some of the most interesting of the nineteenth century designs with mirror are domestic in scale. There we see experimentation with the mirror rather than merely recreating aristocratic themes. A case in point is the London townhouse of Sir John Soane. He eliminated the lavish frames of the past and installed mirrors in varied locations. Often they were small-scale accents. They were inserted in niches behind slender columns. Perhaps only a few inches wide, they might appear between windows. One saw them also on pilasters, and on movable panels. Convex mirrors were applied in unexpected, upper reaches of rooms.

The convex mirror had been popular in England and France in the eighteenth and early nineteenth century because of the way it would provide a diminutive perspective of an entire room. Typically, it was used over a mantel, not in a corner near the ceiling in the Soane fashion.

In the Soane dining room, the mirror is used in conjunction with the mantel, but with a different twist as well. It was not placed directly over the mantel in the traditional French fashion. Instead, a portrait was placed over the mantel. The space at the top and sides of the portrait were mirrored instead. Soane reversed the void/solid relationship. The portrait was treated as a solid, virtually suspended in illusionary space (Fig. 1-13).

Soane wrote about his use of the mirror in his 1835 description of his house and museum. Describing the breakfast room, he noted that it was a limited space but contained varied vistas, both real and illusionary: "The views from this room into the monument court and into the museum, the mirrors on the ceiling and the looking glasses, combined with the variety of outline and general arrangement in the design and decoration of this limited space, present a succession of those fanciful effects which constitute the poetry of architecture" (Fig. 1-14).

Another very small space made memorable by mirror was the bathroom of Madame Beauharnais in her Paris townhouse. Mirrors were installed in small squares, yet they were treated as planes of illusionary space, completely encircling the room. Columns flanking the bathtub were reflected in the mirrors, giving scale to the infinite reflections (Fig. 1-15). Once again we see a shifting of the void/solid relationship. The mirrors were not voids surrounded by solid frames and walls, but planes of space reflecting solids.

Besides experimenting in multiplying views, others experimented in shifting views. In 1816 Humphrey Repton, a landscape architect practicing in England, wrote of the angled mirror:

The position of looking glasses, with respect to the light and cheerfulness of rooms, was not well understood in England during the last century, although on the continent the effect of large mirrors had long been studied in certain places. Great advantage was in some cases taken by placing them obliquely, and in others by placing them opposite: Thus new scenes and unexpected effects were often introduced.

A circumstance occurring by accident has led me to avail myself in many cases of a similar expedient. Having directed a conservatory to be built along a south wall, in a house near Bristol, I was surprised to find that its whole length appeared from the end of the passage in a very different position to that I had proposed; but on examination

Fig. 1-13. Sir John Soane, residence, London, 19th century. View from library into dining room. The overmantel mirror became a surround to the painting. (*Courtesy Trustees Sir John Soane Museum*)

11

Fig. 1-14. Sir John Soane, residence, London, 19th century. View into breakfast room. A small space enlarged with mirrors and windows. (*Courtesy Trustees Sir John Soane Museum*)

Fig. 1-15. Hotel Beauharnais, Paris, 19th century. Mirrors span from wall to wall. (*Courtesy German Embassy, Paris*)

I found that a large looking glass, intended for the salon (which was not quite finished to receive it) had been accidentally placed in the green-house, at an angle of forty-five degrees, showing the conservatory in this manner. And I have since made occasional use of mirrors so placed to introduce views of scenery which could not otherwise be visible from a particular point of view.

For Repton, the mirror now took on another dimension. It was more than a means of expanding space or dematerializing form. It was a way of shifting desirable views, an important function for a landscape architect. Repton was now making a standard part of his design vocabulary the angled mirror, a device that had long been used by magicians as a means of shifting views or making objects disappear on stage.

MAGIC TECHNIQUES

Repton was fooled by the angled mirror, thinking it to be pure space when he first saw it. He was fooled because, due to the angle of the mirror, the light from his face didn't bounce from the mirror's surface back to his eyes, but instead off to the side. Since he didn't see his reflection, the images he did see seemed totally realistic (Fig. 1-17).

Though this is a book about a material as used in architecture, that material sometimes has the curious ability to disappear, seemingly right in front of our eyes. Even in architectural settings, it can be highly illusionary. It seems useful then, in order to understand fully the illusionary aspect of this material, to understand how this characteristic has been exploited.

For magicians, it was especially important that the audience not see their reflections. The angled mirror was useful because of the way it could shift images, or make seemingly solid forms disappear, as the audience just sat there, baffled.

Fig. 1 16. Bathroom of Napoléon III, French, 19th century. An example of multiple reflections and painted mirrors. (*Courtesy French Cultural Services*)

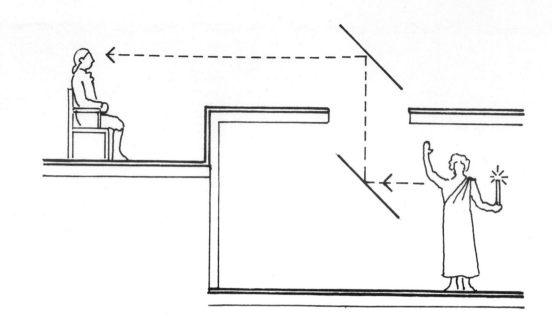

Fig. 1-18. Stage ghosts as produced by angled mirrors.

Fig. 1-17. Repton's mistake. With the mirror at a forty-five degree angle, he thought his contractor had built a new conservatory in the wrong location.

THE TALKING HEAD.

Fig. 1-19. The angled mirror could convey the illusion of complete transparency. From Hopkins, *Magic,* New York, 1901. (*Courtesy Library of Congress*)

14

In one of the earliest recorded descriptions of mirrored stagecraft, *Magia Naturalis* of 1564 by Giovanni Battista della Porta, one learns how "ghosts" were created on stage a hundred years before the Hall of Mirrors. The stage would be dimly lit. Suddenly, a pale spectre would mysteriously glow in front of the audience. The vision was actually an actor standing below the stage. Brightly lighted, his image was projected upward through a trap door by an angled mirror. The image would then be reflected onto a sheet of clear glass, also set at an angle. Since the glass was at an angle, the audience could not see its own image. Since the glass was clear, other members of the cast could walk behind the ghost, making it seem all the more evanescent, a sixteenth century periscope image (Fig. 1-18).

The angled mirror also concealed extraneous elements. A woman who was "beheaded" was actually crouching behind a table. The rest of her body was concealed with an angled mirror. Due to the angle, the audience saw reflections of neutral draperies and carpeting to the side of the stage rather than itself. Since those reflections were identical to the materials at the back of the stage, the effect of complete transparency was achieved (Fig. 1-19).

In an intricate trick involving a woman with seemingly no legs (called "The Living Half of a Woman"), mirrors covered her lower extremities. In order to make the trick seem especially realistic, the magician had nailed table legs to the floor at the sides of the stage. Those table legs would then be reflected in the mirror to look like the otherwise missing fourth table leg. Magicians took great care to create the effect of transparency (Fig. 1-20).

An even more elaborate trick was the magic cabinet. A woman would enter a sort of armoire on stage, surrounded by members of the audience to make sure she did not escape from the back or bottom of the cabinet. The magician would then close the doors of the armoire and then, upon reopening them, discover that his assistant had disappeared completely.

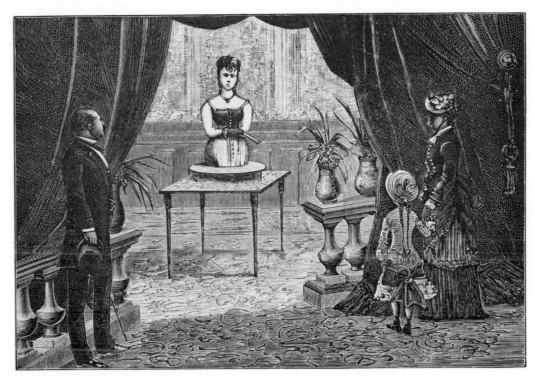

Fig. 1-20. A trick called "The Living Half of a Woman." Her lower extremeties were concealed by angled mirrors. From Hopkins. (*Courtesy Library of Congress*)

15

MAGIC CABINET.

Fig. 1-21. "The Magic Cabinet," from which a woman would completely disappear, all with the aid of mirrors. From Hopkins. (*Courtesy Library of Congress*)

Of course, she didn't really vanish. All she did was to pull some angled mirrored flaps around her while the door was closed. The flaps reflected the sides of the cabinet which, looking like the rear of the cabinet, made the box look empty (Fig. 1-21).

It helped that there was a post in the center of the door to the cabinet to conceal the edge of the glass. All the magic tricks involving mirrors tended to cover the edge of the glass. The mirrors were kept very clean, and the lights dim so as not to have any telltale flashes of light from the mirror's surface.

The angled mirror was not the only mirror type in magic shows. *Glaces à répétitions* were popular in fairs. A triangular room might be totally mirrored. Fairgoers would enter the space through a trap door. Their multiplied images would then resemble a surging mob, particularly if they started to gesture dramatically (Fig. 1-22).

The one-way mirror was also used. We undoubtedly think of this material as being a modern product, particularly used on office buildings and in observation rooms. But it was also used in fairs and expositions at the turn of the century.

Typically, a fairgoer would enter a darkened room and see what looked like a mirror. Then a shutter was opened behind the glass, replacing the mirrorlike effect with one of transparency. The fairgoer's reflection would then be replaced by that of a fiendish devil. The devil was of course another person standing behind the glass, looking particularly disembodied since his shoulders had been covered up with an angled mirror (Fig. 1-23).

Mirrors were also used to create illusionary architectural settings. A "Palace of Mirages" was similar to a cyclorama, but it had changing romantic scenes created by multiple mirrors and painted backdrops. In every direction one could see exotic, infinite vistas. One minute the scene would be Arabian, the next, perhaps Egyptian.

AN OPTICAL ILLUSION PRODUCED WITH THREE MIRRORS.

Fig. 1-22. *Galces à répétitions* were popular at fairs and expositions. From Hopkins. (*Courtesy Library of Congress*)

Fig. 1-23. One-way mirrors were also popular and startling devices for amusement. From Hopkins. (*Courtesy Library of Congress*)

For the magician or manager of a Palace of Mirages, the mirror's illusionary qualities were far more important than its surface qualities. Those illusionary characteristics could be boosted through careful design controls, and a knowledge of where the audience would be in relation to the mirror.

THE AMERICAN SCENE

America was late in utilizing the mirror. Whether in magic shows or parlors, mirrors were quite expensive in America until the twentieth century. Prior to that time, high-quality glass and mirrors were usually imported from England, France or Belgium.

In colonial times, mirrors were highly prized and treated as heirlooms to be passed from generation to generation. At this time, a typical wall mirror would be about 2' x 3', with a decorative but not elaborate frame. It would usually be installed vertically over a couch or table.

In 1761 one Sidney Breese, a New York importer, offered looking glasses (the common term of wall mirrors in those days) for the sum of thirty pounds. One gets the idea of the value of thirty pounds at that time in the following quote by a Lewis Evans of Philadelphia. In 1754 he wrote: "A good farmer's servant is hired by the year in our province at about fifteen or eighteen pounds." So, in the American colonies, a decorative wall mirror was equivalent to the yearly wages of two good servants. Obviously, such a luxury was beyond the reach of the servants themselves.

In spite of the cost, not all experimentation was stifled. Benjamin Franklin exploited the angled mirror to shift views. He designed a small box to be installed outside a second-story window. It contained a series of angled mirrors to give the occupant of the room a view of who was coming down the street or who was standing at the door. The occupant would typically not be observed since the device was unobtrusive, and unexpected. Franklin's invention is sold in hardware stores today under the name of the "Philadelphia Busybody."

With the eventual large-scale manufacturing of mirrors in America, their price would fall to such an extent that they would become everyday items, readily available in dime stores and hardware stores. The material that had once been the prized possession of only the aristocracy or very wealthy would be available to virtually anyone.

MANUFACTURING PROCESSES

As we have seen, the quality of mirrors has been intimately connected with the quality of glass. In America, there had been glass manufacturing since colonial times. In fact, glassmakers came over with Captain John Smith's first colonizing group. But that early venture failed, as did many other American ventures in glassmaking.

Glass was among the last of all industries to become modernized because so much of the production was handled by a few highly skilled craftsmen. In 1900 for instance, window glass was still made by hand-blowing cylinders, then flattening them, much like the Venetian fashion. The resultant glass was wavy and imperfect. Compressed air was substituted for lung power in 1903, but the real technological improvement was flat drawing of window glass, called sheet glass. The sheet glass emerged smooth and bright from its process, but was not perfect enough for mirrors. High-quality plate glass was still reserved for that.

In the 1930's, plate glass could be formed in a continuous ribbon, but still required grinding and polishing. In the 1950's, twin grinding was introduced into the United States. Grinding machines, operating on the top and bottom of plate glass, resulted in near-perfect parallelism.

Today, the expensive grinding of glass has been virtually eliminated by the float process. Molten glass is not cast, but is literally floated on a bed of molten tin, then drawn off in a continuous sheet to the annealing oven. The glass emerges bright and smooth from this process. And today, most glass mirrors are made from float glass. After the glass comes to the mirror manufacturer, considerable care goes into applying the reflective surface. The glass must be thoroughly washed and inspected. It is then sensitized with a solution of tin; though the tin is washed off, it gives the glass an adhesive quality for silvering. The silver is sprayed on the mirror typically, often in as many as 18 coats. It is then protected on the back with a copper coating and a baked enamel finish.

Besides glass mirrors, other materials exist today with the requisite smoothness and brightness to qualify as mirrors. There are a whole host of plastics with metallic finishes. These are lighter than glass mirrors, can be cut and sometimes shaped with ease. Some of these products may have distorted images unless they are applied to a rigid backing.

THE MIRROR AND MODERN DESIGN

In the past 50 years, the term "modern design" has had many different appearances and interpretations. Sometimes the mirror has been an important part of modern design. At other times, It has been virtually ignored. In the United States, when modernist influences from Europe first began to emerge during the 1920's, the mirror was extremely popular.

Modern design ideas began to filter into the United States in the mid-1920's. Although the United States had its modern high-rise architecture in Chicago, and was the home of one of the world's most innovative architects, Frank Lloyd Wright, the modern look most acceptable in the 1920's was art deco. The international style was a revolutionary movement, eschewing extraneous mass and decoration in architecture; it would be some time before it would become firmly entrenched in the United States. Art deco, on the other hand, was a more evolutionary movement, permitting mass and decoration to remain in architecture.

Art deco was an exceedingly eclectic approach to design. It originated in France just prior to World War I. Initially, it was an aristocratic style featuring luxurious materials, woods and elegant hand craftsmanship. Aesthetic influences emanated from African art, Oriental touches in the Bakst stage designs for the Ballets Russes de Monte Carlo and even Viennese *art nouveau.* As the 1920's moved into the 1930's, the style became less aristocratic. Objects were more likely to be mass-produced and were designed ever increasingly for a mass public.

The first art deco ideas to reach America were largely the result of the Exposition of Modern Industrial and Decorative Arts held in Paris in 1925. Though the United States was not represented in the show, numerous representatives of the American press attended. The general reaction was voiced by a writer for *House and Garden* who said, "it is very much doubted if the modernist movement as applied to the home ever gets a strong foothold in this country . . . as a people, we are not ready for such radical innovations as these."

Ready or not, very quickly one saw pictures in American design magazines of fabrics with bold zigzag or geometric designs, geometrically patterned carpets, chrome furniture and numerous mirrors. Mirrors were an integral part of the art deco look. The mirrors would typically be without a frame. If the frame existed at all, it was usually chrome. Sometimes the glass was striated and the overall shape was likely to be faceted or fan-shaped.

American designers took elements of art deco design and added their own touches. The Mayan motif was popular in America in the late 1920's and early 1930's, as demonstrated in the massive looking bathtub surround by R. Harold Zook. The mirror was an important element in the design, artificially lighted and decorated with nautical themes (Fig. 1-24).

Fig. 1-24. R. Harold Zook: house of J.J. Harrington, Glencoe, Illinois, 1930. A massive, Mayan-influenced surround. The mirror was painted with nautical themes.

Another popular motif in American art deco was the New York skyscraper, which itself contained art deco embellishment at its entrance and termination. Though the skyscrapers also had a vague Mayan resemblance, their ziggurat profile was largely due to zoning requirements so that minimal shadows would fall on New York streets. The stairstepped look subsequently appeared in wooden chests, bookcases, even the shape of mirrors. In one intriguing Park Avenue entrance hall, mirrors lined the walls of the space. Then, pictures of little airplanes aloft and skyscrapers under construction were painted directly on the mirrors, resulting in a *glaces à répétitions* celebration of technology.

After the stock market crash of 1929, design became simplified, less stylized and expensive. The faceted mirror was likely to be replaced by the round wall mirror, a seemingly ubiquitous design element in the early 1930's. Another common element at that time was the mirror spanning from wall to wall. The design inspiration for these installations was sometimes Hollywood. In 1935, an advertisement for mirrors in *Architectural Forum* noted how "alert" Hollywood designers were. And a picture of Mae West was included, standing next to a full-length mirror on the set of "Page Miss Glory."

Mirrors were popular in the 1930's in residences and in commercial installations such as shops, beauty salons and restaurants. After the repeal of prohibition in 1933, mirrors found a new home as thousands of restaurants, hotels and clubs underwent remodelings to install new bars and cocktail lounges.

When Prohibition was first instituted after World War I, the bar was typically a dark, oak-paneled saloon, more appropriate for male patrons than female patrons. But by 1933, society had changed considerably. Women were a more visible part of the work force, smoked cigarettes, drove automobiles and could enter bars without being scandalized.

In 1933, an article in *Architectural Record* magazine instructed architects on how best to design bars in the new era of modernity and freedom:

The legalization of alcoholic beverages brings new problems. There is a new age of drinkers. Women cast a decisive vote that can make or break a business. Their entry into business on a large scale has compelled business to (a) establish high standards of sanitation, (b) create more cheerful, artistic and luxurious surroundings, (c) install lighting that glorifies the face, figure and costume, (d) maintain beautiful and convenient toilet facilities, (e) provide comfortable chairs and lounges, (f) furnish stimulating entertainment in modern surroundings.

The article then went on to suggest specific decorative accessories:

Use of more mirrors and decorative plants. Spotlights for dancing. Metal bars, aluminum chairs and composition table tops for novelty bar rooms and modern dining rooms.

Within a few years the new bars were complete, featuring small-scale furniture, light colors and reflective surfaces, completely different from the bars of another era (Fig. 1-26).

The designers in the 1920's and 1930's who used mirrors very often treated them frankly as surfaces. The painted decorations, striated glass and lack of frames all emphasized surface rather than illusionary depth. Typical of this attitude was the flamboyant entrance hall by British architect Raymond McGrath (Fig. 1-27). The design featured striated, curving mirror and smooth wall-to-wall mirror. The latter reflected strong architectural elements, such as curving soffit and stairway. However, one realizes quickly that the

Fig. 1-25. Joseph Urban: apartment of Katherine Brush, New York, 1933. The round mirror was popular in the thirties. This was an updated version of *glaces à répétitions,* with distractingly imperfect glass. Photo Fay S. Lincoln. (*Courtesy Architectural Record*)

21

Fig. 1-26. Russel Wright: bar at the Restaurant Du Relle, New York, 1933. Illustrated are reflective surfaces of copper. Glass mirror occurred behind the bar. Photo Rotan. (*Courtesy Architectural Record*)

Fig. 1-27. Raymond McGrath: St. Anne's Hall, Surrey, England, 1936. Christopher Tunnard's garden plan of the house is sandblasted on the curving mirrored wall. Mirrors were an integral part of the design. Photo Herbert Felton.

sense of depth conveyed in the glass is only an illusion. The doors in the wall, the real voids, are treated as solids. A more traditional approach might have been to apply mirrors to the doors, surrounding them with heavy frames to make them look all the more like voids.

The mirror's popularity was soon to wane. As the world embarked upon World War II, construction not related to the war effort experienced a slowdown. What construction there was utilized few metals since they were diverted for military uses. In America, the work of Frank Lloyd Wright became increasingly popular. Its emphasis on natural materials and strong expressions of the hearth created an appealing image during wartime.

Wright had written about the aesthetic possibilities of the mirror in a 1928 *Architectural Record* article on glass. Yet, he never used it in his own work. Usually confined to the dressing room, it had no spatial importance. For Wright, the concept of space was an unimpeded flow, from one living area to another. The outer wall could be glass, including mitered glass corners, where he deemed it appropriate. In other words, Wright had broken up the mass of the building with open planning and a new freedom in relation to the out-of-doors. He didn't really need an illusionary device to break up the mass of the building further. According to a Wright scholar, William Storrer, Wright used metal mosaics as a means of amplifying light in some of his later projects. But Wright often did not execute the designs, entrusting the task to his secretary Jean Masselink. But for the most part, mirrors had little significance in his work.

After the war, an even more reductive form of architecture came into vogue, the international style. Though it had emerged at the same time as art deco, in America it had remained an avant-garde form for some time. However, in the 1950's, it spread across the country. At this time, there was a tremendous construction boom to fill the need for houses, schools and office buildings that had simply gone unmet during the previous twenty years. The curtain wall was a useful form because it was lightweight, took little time to erect and required less expensive foundations.

Since the mirror is a natural ally of mass in architecture, there was hardly any spot to put a mirror in these new buildings. The few walls that existed on the interiors often didn't even go up to the ceiling. As for the exterior, it was now totally dematerialized, the sheerest possible form of glass skin (Fig. 1-28).

Other factors of taste and philosophy were also operating to make the mirror unwelcome. During the 1950's especially, designers were imbued with a puritanical attitude about materials. A steel building was supposed to look like a steel building, not a marble building or a concrete building. Obviously, a material like the mirror that was deceptive or ambiguous was unlikely to be seen in that intellectual climate. According to Peter Blake, "Everyone was so purist. People felt that anything that was not honest was to be avoided."

An important notion at this time was that the building's exterior was supposed to express the nature of the building's interior. Buildings of Mies van der Rohe were volumes defining neutral spaces inside. Works of Le Corbusier and Frank Lloyd Wright were more plastic, as if the interior spaces were actively pushing the building's exterior form. These buildings were not larger in scale on their outsides with massive attics or outer domes in the manner of traditional architecture. Hence, the converse was also true. Since these buildings did not contain the sort of "illusionary space" that mirrors provide, their interiors were not seemingly larger than their exterior envelopes.

There was also an avoidance of decoration. And mirrors often had the connotation of a decorative, applied material. Mirrors also seemed rather old-fashioned to certain designers. Le Corbusier, in *Vers Une Architecture*, had written of "mirrored wardrobes" as but one of the numerous unnecessary items he wanted to eliminate in his new machine for living. And during the 1950's, designers derided the round mirrors and chrome furniture of the 1930's as representative of "lunchroom modern."

Fig. 1-28. Ludwig Mies van der Rohe: Tugendhat House, Brno, Czechoslovakia, 1930. With the complete dematerialization of the building's mass, there were few locations available for mirrors. Work of the 1950's continued this prototype. (*Courtesy The Museum of Modern Art, New York, The Mies van der Rohe Archives*)

So for a wide range of reasons, the architectural mirror was not in evidence in the middle of the twentieth century in the United States. This is not to say that international style designers positively never used the mirror, or never had used the mirror. A number did, although not to the extent that art deco designers had in the 1930's. Mies van der Rohe used mirrorlike materials, and his earlier Barcelona Pavilion made exquisite use of the reflective qualities of dark marble and pools of water. Le Corbusier, though he wrote negatively of the mirror, actually installed it in a few of his designs in a small scale way. A well-known living room designed by Le Corbusier and Pierre Jeanneret (Annex to Church Villa, Ville D'Avray, 1929) had a mirror built into one corner; the installation is so illusionary that this writer had looked at the picture on numerous occasions before noticing the existence of a mirror.*

Richard Neutra was one of the few international style designers for whom the mirror was an integral part of his work. He treated mirrors as if they were planes of clear glass, and juxtaposed them with large sheets of clear glass. The effect was often startling and confusing.

Peter Blake used mirrors in exhibit designs as one way of expressing the seemingly limitless quality of art and architecture in America at this time. In 1949, he designed a model for an exhibition space to house works by Jackson Pollock. Pollock's paintings were to be treated as planes in an open, Miesien pavilion. Perpendicular to the paintings were little spur walls, which had mirrors applied to them. Since the mirrors then were reflecting the painting, as well as other mirrors, one would see the Pollock swirls of color progressing to infinity. "Pollock was fascinated by that," according to Blake. In 1957, Blake used a similar

*p. 127, *The International Style.*

24

approach in an exhibit held in West Berlin. Entitled "America Builds," the exhibit was to demonstrate the latest American building techniques and the latest curtain wall buildings (Fig. 1-29). Full-scale mockups of wall panels were installed in a 28' high exhibition hall. In order to make the mockups seem even more realistic, Blake installed mirrors adjacent to them. As in the Pollock exhibit idea, the mirrors were placed face to face. If the buildings were essentially horizontal, the mirrors would be placed at the sides so that the building would look even more horizontal. For the vertical building, the mirrors would occur at the bottom and top of the panel, giving a *glaces à répétitions* sense of infinite height. Ironically, mirrors were used to convey the essence of an architecture in which illusions were the exception rather than the rule.

Mirrors began to re-emerge in the 1960's as an important architectural tool. One of the first significant interiors was the I. Miller Shoe Salon, remodeled by Victor Lundy (Fig. 1-30). The mirrors in that design were decorative and highly illusionary. They were an integral part of the store's design, spanning the entire back wall and being used in inserts between the columns. The columns themselves were "fake," their scale being enlarged through the use of curving wood screens. Said Lundy of the design: "It is make-believe . . . fun. It was created out of nothingness. Commercial interiors it seems to be are a make-believe process. Don't take it too seriously, just go in and enjoy it." His playful attitude represented a shift in thinking at the time. For Lundy, the mirror was a necessary component in expanding the space of the store. The real confines of the walls were not sufficient.

Lundy was using the mirror in a traditional, decorative fashion. However, a few years after the I. Miller project, a number of young avant-garde architects and architecture

Fig. 1-29. Peter Blake: exhibition "America Builds," West Berlin, 1957. Mirrors were placed adjacent to mock-ups of curtain wall facades to convey a sense of their full dimensions.

25

Fig. 1-30. Victor Lundy: I. Miller Shoe Salon, New York, 1962. A decorative use of mirrors. Photo George Cserna.

students used the mirror in a nontraditional, often highly undecorative fashion. Even though technology had progressed in 300 years so that perfectly smooth and flat mirrors were available, these designers purposely chose warped mirrors with distorted surfaces.

An influential book during the middle and late 1960's was Robert Venturi's *Complexity and Contradiction in Architecture.* The book was a critique of the international style's reductiveness and simplicity, and a demonstration that great works of architecture in the past had richness, multiple layers of meaning, and often tension resulting from the juxtaposition of different systems. The catchword arising from his theories was ambiguity. Architecture of the international style was not ambiguous, whereas ambiguity was a component of many great works in the past, according to Venturi.

Designers influenced by Venturi wanted to incorporate ambiguity into their work. Though Venturi didn't advocate using mirrors, for many designers, mirrored surfaces were a shortcut toward that ambiguity. They chose mirrored wallpaper, silver insulation and mylar. In essence, they were warped, "funhouse" mirrors, not smoothly decorative mirrors.

The mirror itself is a naturally ambiguous material in most cases. If one indeed recognizes it is a mirror, it is a shiny surface that also conveys the illusion of depth. The young designers in the 1960's heightened the mirror's natural ambiguity. With a warped and distored surface, one would know for sure that one was observing a mirror. Hence, one would be unable to avoid the knowledge that it was ambiguous, two things at once, both surface and depth.

Since the images were distorted, the observer would be disoriented. He would no longer move in a safe, rectilinear, modern New York apartment, say, but a bizarre, barely comprehensible world. With curious supergraphics and strobe lighting, also possibly under the influence of drugs, one could feel "spaced out." C. Ray Smith described the attitudes of designers at that time:

The new aesthetic happening evidences an interest in spatial ambiguities, confusion and distortion. What is of interest is a nebulous area between precise perception and the hallucination of infinity. It corresponds to current psychedelic investigations, perhaps by aiming at expanded spaciousness.

Peter Hoppner, a young designer experimenting with warped mirrored surfaces, also attempted to define the period. He said:

The whole spirit of today is like the spirit of the twenties. And people are using silver because they are tired of all white and natural colors. Also, architects are gradually getting interested in the same horrible colors and textures that sculptors are doing today—as opposed to the old purist business. It is in the air.

Hoppner likened the spirit of the mid-1960's to that of the 1920's. However, the attitudes in those times were quite different. In the 1920's and 1930's, the mirror was a symbol of a shining, brave new technology. In the 1960's, designers were bored with the technology that they had known. The clean, all-white, unadorned apartments with large glass windows that had been the dream of the avant-garde in the 1920's and 1930's were now being destroyed visually with a warped funhouse mirror (Fig. 1-31).

RELAXED TASTES

The experiments of the 1960's relaxed architectural tastes. By the 1970's, supergraphics, mirrors and *trompe l'oeil* devices were no longer the province of a small avant-garde, but had become incorporated into the general architectural vocabulary. The warped mirror of the 1960's was replaced by a smooth, frankly decorative mirror. The small modern apartment now would not be "destroyed" with a warped mirror. Rather, it would more likely be expanded with a mirror sometimes installed in highly illusionary fashion. When warped mirrors were used, they were likely to occur in the popular discotheques to add sparkle to those dynamic environments.

Architects were now relaxed about incorporating historical elements into their work. They might install neo-classical moldings and trim in places where previously no ornament was allowed. Many designers reached back to the work of the 1930's for inspiration. The mirrors, light woods and bent chrome furniture of that time were reinterpreted by designers in the 1970's.

Architects also incorporated a jazzy, futuristic-looking reinterpretation of outer space technology in their work. This might result in a mirrored ceiling with bright stainless steel ducts shooting into the mirror's illusionary void. A mirrored wall at Hellmuth, Obata and Kassabaum's Greenwich Savings Bank had cylindrical stainless steel benches cantilevered from the wall. The effect was not of permanence, but of little spaceships about ready to hover away.

This widespread use of the mirror in the 1970's differed from the 1930's, the other time when mirrors had been quite popular. In the 1970's, it was simply far more ubiquitous than had been the case in the 1930's. Serge Chermaeff had used mirrors at that time. Along with Welles Coates in England, he had installed them in shops as a means of making cramped spaces seem wider. But he noted, "Of course, we would never use them in a serious functional environment, or in a residence." For Chermaeff, the mirror was simply not an appropriate material for all locations. In the 1930's, it simply would not have appeared in all the banks and offices that are often home for today's architectural mirror.

Fig. 1-31. Peter Hoppner: distorted pier mirror, own apartment, New York, 1966. Warped mirrors were more ambiguous than smooth mirrors. Photo Louis Reens.

THE NEW NARCISSISM

For every design decision made by an architect, there has to be a client who accepts it and pays for it. By the 1970's, architectural tastes were relaxed so that architects could install a material that they considered ambiguous or illusionary. On the other hand, that material had to be accepted by a client who undoubtedly thought of it as narcissistic. For instance, an architect might wish to install mirrors on the closet doors of a bedroom, realizing the spatial characteristics of the room would be enhanced. The client on the other hand would think about the mirrors and ask himself or herself, "Do I really want to look at myself in those mirrors?" The answer in many cases was obviously yes, due to the ubiquitousness of the mirror in the widest possible variety of interiors in the 1970's.

The widespread use of the mirror in the 1970's coincided with a new narcissistic wave in society. Tom Wolfe talked about it in *The Me Decade*, as did Christopher Lasch in *The Culture of Narcissism*. Conferences were held to discuss it, like one at the University of Michigan entitled "Narcissism in Modern Society."

The narcissistic personality was defined as having an overwhelming interest in himself or herself. It wasn't that the self was particularly wonderful, but at least that the self could be improved and changed and molded. The scale of life seemed so out of the average person's control that that person turned inward. Politics often seemed remote to the individual, and ceased to gain his or her attention. Indeed, there was a significant decline in voter participation in national elections, starting in the mid-1960's. The family was not the bulwark of strength it had been; people were often divorced, and living apart from relatives. Even the future seemed out of control. People did not save for it, preferring to take their money and spend it on present consumption. Banks noted an alarming drop in the savings rate in the 1970's.

Herbert Henden, a psychoanalyst quoted in *Time* magazine, summed up the malaise. "When I grew up, there was a greed for material things; now it's a very egocentric greed for experience." Many of these experiences were mirrored. All the health spas that sprang up in the 1970's were mirrored, as were the sex clubs and massage parlors that reportedly were on the increase. Even tamer experiences occurring in banks and offices were now mirrored.

Charles Gwathmey installed mirrors on some free-standing partitions in an office. At first, he was worried that the clerks and typists working near the partitions would find their reflections distracting. But, according to Gwathmey, he received no complaints.

One of the most intriguing mirrors was installed by Gunnar Birkerts at the Calvary Baptist Church in Detroit. The entire wall over the altar was a canted mirror. The idea behind that installation was for the congregation to gain a greater sense of participation in the ceremony through the group reflection.

While some might argue that the Calvary Church mirror was not narcissistic since it was catering to a group rather than individuals, it is interesting nonetheless. For the congregation didn't mind the idea of looking at themselves while in church. They were not gazing upward at the image of a spiritual Christ, but at the image of temporal humanity (Fig. 1-32).

So there the mirror was, reflecting peoples' images back to them in the widest possible variety of activities. Of course, a traditional location for the mirror, the foyer or place of transition, was not overlooked. In 1970, the Architectural League of New York held a show which

Fig. 1-32. Gunnar Birkerts and Associates: Calvary Baptist Church, Detroit, 1977. An unusual location for a mirror.

29

featured the Miami Beach hotels of Morris Lapidus. Lapidus was an architect previously ignored by his profession. But since his work was so consistently admired by the general public, which was not always the case of some of his more illustrious peers, the League decided to investigate the reasons for his success.

The show demonstrated that Lapidus placed human needs and desires before architectural abstractions. In his hotels, he made a vacationer feel that he or she was a glamorous star, about to make a dramatic entrance. Lapidus would often have a circular staircase in the hotel lobby so that a woman could descend a flight, feeling every bit a Scarlett O'Hara. The stair rail would be wide so that the rings on her outstretched hands would be visible. And at the entrance to the restaurant there would also be a few steps leading down to the throng below.

With all these grand entrances, Lapidus realized that patrons would be especially mindful of their personal appearance. He said, "The first thing a man does before he walks 'on stage' is to check his fly. And the woman checks her hair, the length of her dress, sees if the hem is straight. There's always a mirror there. I'm thinking of coming and going."

In a similar vein, an architectural consultant, quoted in *The Washington Post*, said that a major hotel chain had contacted him because hotel guests were complaining about slow elevator service. The consultant weighed the costs involved with speeding up service. He finally came up with the cheapest solution of all. He suggested installing mirrors in all the elevator lobbies. When that was done, the complaints ceased.

REMODELINGS

Most architects practicing today were trained in the belief that they would spend most of their lives designing totally new, free-standing buildings. However, for a number of designers, this belief has not matched reality. Today, a number of architects have been involved in remodeling older buildings. Partly, this has been caused by the preservation movement that began in the late 1960's. Citizens and writers and designers realized that older buildings imparted a necessary balance to a given neighborhood. The motivations were also economic. With increasing costs of labor and material, it was often cheaper to remodel older buildings than to build anew.

What is interesting about this fact is that these architects remodeling these older buildings were trained in the belief that open planning and abundant glass on the outer wall were necessary components in a well-designed environment. But of course, these older buildings contained solid outer walls, often solid inner walls, and sometimes not very much glass. In other words, they were examples of architectural mass.

Many designers confronting these older structures applied their modern value systems to them. Their knee-jerk reaction was that these structures had to be "opened up." A new meaning for an old verb, to gut, came into being to describe the approach. Floors were removed, bridges and balconies installed in two-story spaces, and holes punched in bearing walls. Traditional buildings were made to conform to newer design tastes.

However, not all remodelings could be as open as the practitioners desired. A small apartment, a basement restaurant, typically had to contend with their existing dimensions. A hole in the wall might just lead to someone else's private domain. In certain cases, illusionary space made a good deal more sense than real space. When open planning and glass walls could not demolish architectural mass, the mirror was able to do it with illusions.

In the 1970's, architects Gwathmey-Siegel began to execute a lot of remodelings. Suddenly, they started using vast quantities of mirrors in their work, because the apartments and offices and restaurants they were remodeling were fairly rigidly circumscribed. As Gwathmey put it, "when the container is given, you have to rethink the nature of space."

Gwathmey considers Le Corbusier his greatest architectural inspiration. Yet, as has been noted, Le Corbusier used mirrors quite rarely. But then, Le Corbusier's *oeuvre* consisted essentially of new free-standing buildings. They could be as open as money and the client permitted, and had abundant opportunities for admitting daylight to the interior. He didn't really need mirrors. But today many architects feel that they do.

THE ENERGY CRISIS

In the 1970's, energy efficiency became far more of a factor in buildings than had been the case in the past. With increases in energy prices that began with the Arab oil embargo of 1973, architects and engineers realized that they could no longer take the thermostat for granted.

The outer wall began to receive a considerable amount of attention due to the poor insulating qualities of clear glass. A beautiful view was not sufficient reason to install a large glass window. Was that window facing north or south? Was heat loss likely to be a greater problem, or heat gain? Should the window be double glazed? Coated with a heat-reflective or heat-absorptive finish? In a word, the essence of modern architecture for so long had been glass. Even when architects began to introduce large massive walls in their designs in the mid-1960's, they knew that they could have vast quantities of glass wherever they needed it. Paul Rudolph's Art and Architecture School in New Haven had been an expression of massive corner towers, with extensive glass infills.

After the energy crisis, some architects began to use less glass in their energy-efficient new buildings. A residence might be open to the south, but with few windows on the north. An office building might now be a totally tinted glass structure, with only a small percentage of the outer

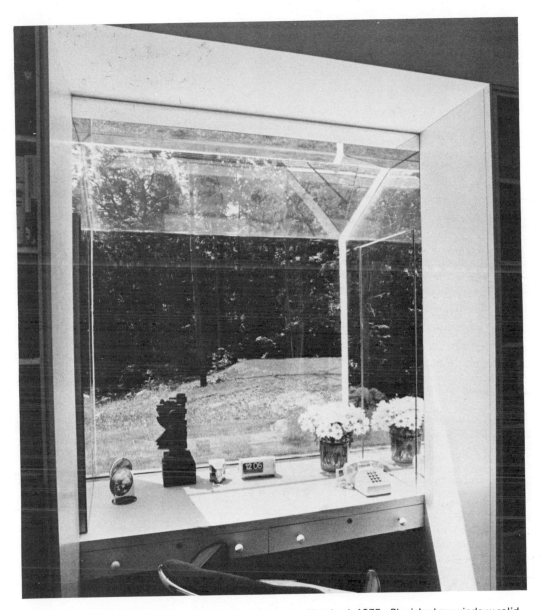

Fig. 1-33. Hugh Newell Jacobsen: Baker House, Frederick, Maryland, 1975. Plexiglas bay window; solid side supports are dematerialized with mirror.

31

wall of the building "vision" glass. The rest would be insulated sandwich panels, looking just like the vision glass due to the similarity of the tint. So from the outside, the building would look like a curtain wall expression of glass. But from the interior, the occupants would know exactly where that vision glass stopped.

In the past, mirrors were used in conjunction with windows in order to amplify views and light, and diminish the mass of intervening piers. One can see opportunities for the same type of application today. In this theme is an interesting window designed by Hugh Jacobsen (Fig. 1-33). Designed more for aesthetic reasons than energy ones, it nevertheless shows how mirrors can be substitutes for glass. Jacobsen mirrored the solid side walls that supported a plexiglas bay window. The mass of the solid side walls disappeared completely due to the illusions of forest that played across the mirrored surfaces. In a sense, the mirrors merely made the plexiglas more efficient by duplicating the views of the exterior.

Another form of efficiency can come through the re-direction of the sun's light. An interesting example of this is the Pragtree Farm Greenhouse, located in the Pacific Northwest (Fig. 1-34). The north wall of the greenhouse was built in a parabolic shape and covered with a reflective material. The sunlight entering the south facing glass would be intensified by the parabolic mirror, helping to warm a massive pool of water, the primary heat source for the greenhouse.

At the outset of this book, it was noted that one could think of the mirror in two ways. One would be a highly illusionary approach, the other would treat the mirror as a straightforward surface, a surface that might shift rays of light brightly. One would be essentially invisible, the other would be at the forefront of our consciousness. In

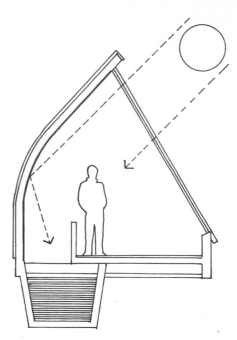

Fig. 1-34. Howard Reichmuth and Jeffrey Barnes: Pragtree Farm Greenhouse, near Seattle. Back wall is covered with reflective material to intensify energy striking water container.

these two windows, we see that the Jacobsen approach favors illusionism. The Pragtree Farm approach treats the mirror more as a surface, in this case a highly functional surface. They are both examples of the extraordinary versatility inherent in the mirror. It can be frivolous or ambiguous, or totally disappear from view, or technical, all because it can reflect light more perfectly than any other material.

2
The Mirrored Surface

*I want people to know it is a mirror. There is
no intent to fool. There is an intent to enrich.*
Charles Gwathmey, 1978

We are generally not fooled by this mirror's illusionary qualities. We know that we are looking at a mere surface, but a surface with illusionary depth. Perhaps this mirror is located so that we can easily see our reflections in it; or its surface might be striated and embellished; it might reflect the flash of lights; or it might be located on and around objects, such as on tables and columns. Since we are aware of the dual nature of this mirror, it is an ambiguous mirror. It is surface and depth simultaneously.

Designers in the 1960's attempted to heighten the mirror's natural ambiguity through surface distortion. Peter Hoppner's crinkled foil hallway typified this design approach. Through an anarchistic gesture, it was remarkably similar in effect to present-day discotheques which are hardly symbols of societal discontent. Both Hoppner's hallway and Regine's are examples of "scattered reflection." If a mirrored surface is bent or fragmented, each portion of the surface will act as a tiny mirror. For Hoppner's hallway, the idea was to create distorted illusions on the surface. At Regine's, the surface deflects light rays in a kinetic fashion. (Figs. 2-1 and 2-2).

Warped or fragmented mirrored surfaces are fairly obvious examples of the mirror as surface. A more subtle

approach would be the way designers treated the mirror in the 1930's. It conveyed illusions and was smooth, but it was frequently an object without a frame applied to the wall rather than seemingly recessed in the wall. In traditional designs, the heavy frame helped make the mirror seem like a void surrounded by solids. The traditional frame also created a visual break between the mirror and what it was reflecting. That way, the slightly greenish tint of the glass was not so noticeable. But designers in the 1930's treated the mirror as something of an appliqué.

We see a similar attitude in some of the recent designs evocative of the 1930's. Robert A.M. Stern's 1979 Westchester County house had a true art deco dressing table in one of the bedrooms, looking right at home with the other curving and decorated motifs (Fig. 2-3). A popular restaurant in New York, Ruskay's, made abundant use of circular and semicircular 1930's-style mirrors, as well as chromium lunchroom type counter, stools and booths. Ruskay's is a romantic interpretation of the 1930's, which means that in many ways it is quite different from that era. For one thing, the quantity of mirrors has increased, covering the top of the counter and tables as well. The walls are painted dark hues instead of the light colors more standard in 1930's-

Fig. 2-1. Peter Hoppner: Hallway, own apartment, New York, 1966.

Fig. 2-2. Justin Henshell: Regine's, New York, 1976. (*Photo courtesy Tomorrow Designs*)

type lunchrooms. Perhaps the most striking aspect of the design is the lighting. Ruskay's is filled with tall, tapering white candles that appear on the tables, and even on the counter. Hence, the hurried, solitary cup of coffee becomes a surprisingly elegant and restful interlude as one sees the candles reflected in all the mirrored surfaces (Figs. 2-4 and 2-5).

The works of Gwathmey-Siegel are also evocative of the 1930's with their large expanses of mirror, curving glass walls, bent chrome furniture, light colored finishes and pale woods. As has been noted, Gwathmey himself admits to being influenced primarily by Le Corbusier, though Le Corbusier never used mirrors in such a widespread fashion.

Pearls restaurant was Gwathmey-Siegel's first major use of mirror. It is an interesting work, because it makes one think of various aspects of the mirror, spatial and psychological, and the relationship of the mirror to the philosophical underpinnings of modern design (Figs. 2-6 and 2-7).

The dining room was an unfortunate proportion initially, only 14' wide, and 100' long. By placing the mirror on the long side of the room, the width of the restaurant suddenly seemed to double. And since it reflected the ceiling, the semicircular ceiling became an illusionary barrel vault.

The architects treated the mirror quite frankly as a surface. So, though it was above eye level and expanding the space in an illusionary fashion, it was easily recognized as a mirror due to a critical design detail. The architects left the edge of the glass exposed as a frank expression of the mirror's nature. According to Gwathmey, the mirror made the space "ambiguous and much more expansive" (Fig. 2-8).

In a way, Pearls is a textbook example of how to use a mirror. If a space is too narrow, the logical place to install the mirror is on the side wall so as to make the space seem wider. This the architects did.

Fig. 2-3. Robert A. M. Stern, Architect: Residence, Westchester County, New York, 1976.

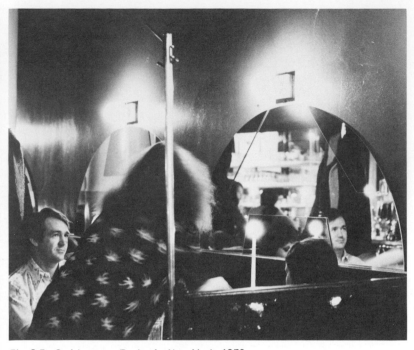

Fig. 2-5. Carl Laanes: Ruskay's, New York, 1973.

Fig. 2-4. Carl Laanes: Ruskay's, New York, 1973.

36

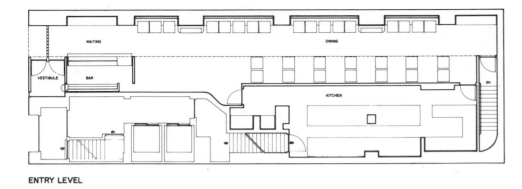

ENTRY LEVEL

CELLAR

PEARLS ⊕

Fig. 2-6. Gwathmey-Siegel: Pearls restaurant, New York, 1973. Plan.

Fig. 2-7. Gwathmey-Siegel: Pearls restaurant, New York, 1973. Section.

37

However, after Pearls opened, some of the patrons complained when seated at the back of the restaurant. They felt isolated. So, the architects installed mirror at the back of the space which had the effect of bringing the brightly lighted entry space to the rear visually. Hence, patrons no longer felt they were seated in Siberia. Instead, they felt right in the middle of things. But, spatially, the restaurant now seemed longer than ever (Fig. 2-9).

As has been noted, Gwathmey was influenced by Le Corbusier. An important philosophical tenet which he carried over from Le Corbusier was the idea that the exterior of the building should reflect what was happening on the interior of the building.

However, if the building's interior contains a large illusionary component, how is that demonstrated on the outside of the building? For it is possible that the interior of a building could seem quite a bit larger than its exterior envelope. Gwathmey-Siegel resolved this dilemma by extending the mirrored wall from the interior of the restaurant to the exterior of the restaurant. Thus, persons passing by could glimpse the illusionary characteristics beyond (Fig. 2-10).

After Pearls, Gwathmey-Siegel continued using the mirror in a large-scale way. Gwathmey considered the mirror as a means of enriching and enlivening space. In a lavish New York apartment they installed mirrors in the children's playroom, even in the kitchen. When questioned about the kitchen mirror, Gwathmey replied, "The kitchen is a legitimate space" (Figs. 2-11 and 2-12). The kitchen mirror was placed on an axis with a door to the dining room. So, when that door was opened, one would see the reflection of the dining room rather than the kitchen per se.

Fig. 2-8. Gwathmey-Siegel: Pearls restaurant, New York, 1973. A completion of an architectural form with mirror. Photo Bill Maris.

Fig. 2-9. Gwathmey-Siegel: Pearls restaurant, New York, 1973. Eventually, the rear of the space was mirrored. Photo Bill Maris.

Fig. 2-10. Gwathmey-Siegel: Pearls restaurant, New York, 1973. The interior illusionary space is larger than the exterior physical envelope. Photo Bill Maris.

Fig. 2-11. Gwathmey-Siegel: Swid apartment, New York, 1978. Colors and materials are reminiscent of the 1930's. Photo Norman McGrath.

Fig. 2-12. Gwathmey-Siegel: Swid apartment, New York, 1978. Why not mirrors in the kitchen? Photo Norman McGrath.

Robert A.M. Stern used a mirror at the Erbun Fabrics showroom in New York as a means of concealing a storage area while at the same time revealing in a kaleidoscopic fashion the brightly colored fabrics on display. According to Stern, the rectilinear lattice work on the mirror was an allusion to the work of Charles Rennie MacKintosh. The lattice, as well as the diagonal disposition of the mirror, gave the surface a subtle effect (Figs. 2-13 and 2-14).

Roche/Dinkeloo created an elegant and opulent swimming pool atop the U.N. Plaza Hotel and Office Building in New York. Mirrors surrounded the columns in the swimming pool, reflected the horizontal banding of the windows, and reiterated that horizontality with joint placement. The dramatically tented ceiling provided a foil to the shininess of the pool, mirror and glass (Fig. 2-15).

Roche used mirrors on the columns at the U.N. Plaza and on tables and counters at Richardson-Merrell. However, the sort of mirror chosen at Richardson-Merrell was mirror glass, partially reflective, partially transparent glass. Hence, the tables and counters are more subtle than would be the case with regular mirror (Fig. 2-16, Plate 1).

In Roche/Dinkeloo's addition to John Deere, they installed mirror glass on an interior partition in the cafeteria. For the most part, the wall functioned as a regular mirror; it reflected the interior garden in the cafeteria. However, at one point, a tapestry placed immediately behind the glass was spotlighted, making the glass appear transparent in that location. As the work of art emerged from the reflections, one questioned which was more real, the reflections or the tapestry (Plate 2).

Though most architects think of mirror glass as primarily an exterior material, for Roche/Dinkeloo, it has assumed important decorative dimensions on the interior as well.

Fig. 2-13. Robert A.M. Stern: Erbun Fabrics, New York, 1979. A mirror to reflect multicolored fabrics. Photo Ed Stoecklein.

Fig. 2-15. Kevin Roche, John Dinkeloo and Associates: United Nations Plaza Hotel, New York, 1975. The mirrors dissolve the columns and reflect view of East River and United Nations across the street.

Fig. 2-14. Robert A.M. Stern: Erbun Fabrics, New York, 1979. Photo Ed Stoecklein.

Some designers exploit the mirrorlike qualities of other materials (Fig. 2-17). Warren Platner has experimented with the reflective qualities of clear glass (installing it on the hung ceiling and on top of a tile floor at the Steelcase Chicago Showroom) and the reflectivity of highly polished metal. His elegant bookshelf and lighting fixture at his residence are polished metal (Figs. 2-18 and 2-19).

For John Portman, mirrors have been a necessary ingredient in the design of his hotels. In the Detroit Plaza Hotel at Renaissance Center, a mirrored partition both conceals and reflects, in a highly fragmented fashion, the exterior view. Portman is careful to juxtapose his mirror with other surfaces of contrasting texture. This is particularly evident in the Terrace Room at his Peachtree Plaza Hotel in Atlanta. And in his various hotel skytop dining rooms, mirrors on walls and ceiling celebrate surrounding views (Figs. 2-20, 2-21 and 2-22).

Fig. 2-16. Kevin Roche, John Dinkeloo and Associates: Richardson-Merrell Inc. Headquarters, Wilton, Conn., 1974. One-way mirror used on tables.

Fig. 2-17. Warren Platner Associates: Steelcase Chicago Showroom, Chicago, 1969. Clear glass installed so as to be reflective rather than transparent. Photo Ezra Stoller.

Fig. 2-18. Warren Platner Associates: Platner residence, New Haven, Conn., 1971. Photo Ezra Stoller.

Fig. 2-19. Warren Platner Associates: Platner residence, New Haven, Conn., 1971. Photo Ezra Stoller.

Fig. 2-20. John Portman and Associates: Renaissance Center, Detroit, 1977. Mirror that both screens and reflects. Photo Alexandre Georges.

Fig. 2-21. John Portman and Associates: Peachtree Plaza Hotel, Atlanta, 1977. A striated mirror reflecting varied, contrasting textures. Photo Alexandre Georges.

Fig. 2-22. John Portman and Associates: Peachtree Plaza Hotel, Atlanta, 1977. Mirrored ceilings amplify the view.

Fig. 2-23. Gwathmey-Siegel: Shezan restaurant, New York, 1976. Industrial acoustical ceiling tiles convey a lightness and sparkle to this basement space. Photo Norman McGrath.

Mirrored ceilings were virtually a hallmark of the 1970's, partially due to the introduction of some new mirrored products. One was a lightweight acoustical tile with a stretched, mirrored plastic facing. Another was an acoustical tile with a spectral aluminum facing.

In the past, mirrors attached to the ceiling had to be bolted to the structural system, and still contained some element of danger. In a popular movie entitled "Heaven Can Wait," Warren Beatty barely missed sudden death when a mirrored ceiling crashed down on a bed he was about to occupy. And in Washington, D.C., a large mirror company refuses to install regular mirror ceilings because of the risk involved. But with the new lightweight mirrored ceilings, the most damage that could occur form a falling tile would be a spilled carafe of wine.

At Shezan restaurant, Gwathmey-Siegel installed the aluminum acoustical ceiling tiles as a means of disguising the fact that the restaurant was underground. Since the massive columns in the space penetrate the ceiling, they appear to be taller than they actually are. The whole effect, according to Gwathmey, was to be like a giant, lightweight tent (Figs. 2-23 and 2-24, Plates 4 and 5).

That particular ceiling tile had originally been used in kitchens and swimming pools. Gwathmey-Siegel were among the first architects to realize its aesthetic quality.

Gwathmey described the design approach:

Fig. 2-24. Gwathmey-Siegel: Shezan restaurant, New York, 1976. Photo Norman McGrath.

47

Fig. 2-25. Blake Millar: Noodles restaurant, Toronto, 1973. This restaurant has a decided outer space imagery. The ducts appear to shoot into the illusionary void. Photo Robert Perron.

The obvious strategy was to create an environment of illusion and elegance, where the perceptual space is simultaneously real and ambiguous. The use of reflective materials (mirrors on the walls, polished aluminum on the ceiling) and glass block, contrasted to the beige travertine floor, oak banquettes and grey carpeted walls, establishes the physical reference. The spaces become at once subdued and crystal, defined and infinite, crisp and soft due to the juxtaposition between opacity and reflectivity.

The total effect at Shezan is one of calm serenity. In Noodles, another restaurant with a mirrored ceiling, the effect is more complex and evocative of outer space technology. Noodles, a Toronto restaurant remodeled by Canadian architect Blake Millar, had to contend with existing ceiling ducts. Millar mirrored the ceiling, covering the bothersome ducts as well as new ducts over a cooking island with annealed stainless steel. The ducts seemed to penetrate the illusionary space created by the glass ceiling, almost looking like space vehicles (Figs. 2-25 and 2-26).

Millar wanted Noodles to be different from the usual Toronto restaurant. He said he wanted it to be a "glistening, reflective salon rather than another restaurant with fake beam ceilings and red banquettes."

One intriguing aspect of Noodles was that one of the most desirable seating areas in the house was the mezzanine. It was treated as a sumptuous, comfortable perch from which to view the comings and goings and food preparation below. Since the mirrored ceiling reflected the activity below, one did not feel isolated in the area.

Of course, mirrors on ceilings do not have to be applied in large, smooth sheets. There exists the tradition of small bits of mirror, applied to ceilings and walls in the form of mosaic. Evidently, the mirrored mosaics in Iran were a result of the arrival of mirrors from Europe which broke in transit. The Iranians made use of the mirrors anyway, creating a new aesthetic form.

Fig. 2-26. Blake Millar: Noodles restaurant, Toronto, 1973.

49

In a similar vein, the mirrors applied on the walls and ceiling of artist Isaih Zagar have a distinctly free-form quality. He smashed the mirror and applied it on the ceiling and walls of his apartment. The assemblage has a primitive quality about it; it is hard to believe that it is in the middle of Philadelphia (Fig. 2-27).

We know the mirror in Zagar's apartment is a surface. Yet, because of its illusionary qualities, it dematerializes mass and amplifies existing light in the space. While accomplishing these tasks, because of the way the designer used it, this mirror looks totally different from other examples of the mirrored surface.

Fig. 2-27. Isaih Zagar: Eyes Gallery and private residence, Philadelphia, c. 1976 (project is ongoing). Wire lathe and plaster cement in which are embedded broken pieces of mirror.

3
Deceptive Mirrors

A mirror is a mirror when you can see yourself.
Otherwise it is an illusionary device for expanding space.

Hugh Newell Jacobsen, 1979

The essence of the mirror is its ability to reflect light perfectly. Hence, it can create illusions on its surface. It's just that sometimes we are completely fooled by those illusions into thinking that the mirror simply isn't there. It ceases to exist on a perceptual level.

Usually, we think of illusionary installations as being approximations of pure space. This was the basis of the magician's angled mirror. Also, most of us have undoubtedly attempted to walk through a mirror by accident. An acquaintance with a lavishly mirrored apartment said that visitors to the apartment were always bumping into the mirrors.

Sometimes when mirrors are placed perpendicular to glass windows or in transom areas they resemble glass, separating indoors from outdoors or one space from another. In these cases, the surface of the glass is visible, it's just that we are not necessarily aware that there is a reflective coating on the back of the glass. In a sense, this is a new way of being fooled by mirrors. We are so used to glass windows and glass partitions in modern design, that we are easily fooled by something that just looks like clear glass.

In a rather amusing example, a photographer was dispatched to photograph some mirrors used in an unusual way in an office. The firm of Mills, Clagett and Wening had installed glass and mirror clerestories above partitions in a modern office design. In certain instances, ducts penetrated below the ceiling. The architects decided to disguise the ducts by making some of the clerestories mirror instead of glass. The pattern was random, depending on where the ducts occurred.

When the photographs of the "mirrored" clerestories were developed, one could see that the photographer had been fooled by the mirrors. Not knowing exactly which was which, he had photographed glass clerestories instead. When examining the photographs carefully, one could see the mistake (Fig. 3-1).

APPROXIMATIONS OF SPACE

The magician with angled mirrors made solid forms disappear visually. Don Clay, a New York designer, performed a similar feat by mirroring the bottom of his custom-designed platform bed. Though the mirror was not angled, it was positioned so that one would be unlikely to see the reflections of one's feet in the mirrors (Fig. 3-2).

For Clay, the mirror covered a solid form. For Gamal El-Zoghby, another New York designer, the mirror is treated as a void in a solid wall. It is a window into another realm. El-Zoghby describes his techniques and reason for using mirrors:

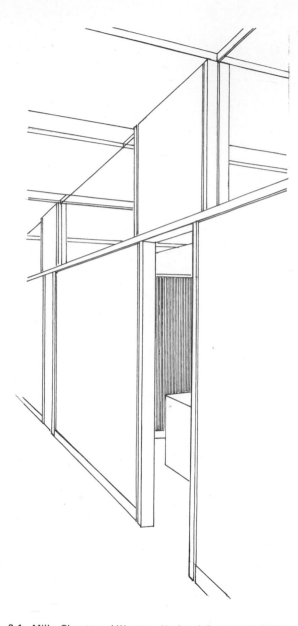

Fig. 3-1. Mills, Clagett and Wening. National Geographic Society Computer Facility, Darnestown, Maryland, 1979. Glass and mirror transoms are used interchangeably.

Fig. 3-2. Don Clay: own apartment, New York, 1979. With a mirrored base, the bed seems suspended in space. Photo Peter M. Fine.

Mirrors in my designs are neither bevel-edged, framed nor jointed. They have to be recessed to deny the mirror as an object of reality rather than to express the illusionary reflectivity as a subject. In the continually diminishing spaces for urban dwellers, my use of mirrors has served the psychological function of overcoming the effect of confined spaces.

Along with the recessed mirrors, El-Zoghby creates a series of real spatial voids—a balcony, a sitting ledge, a window into the kitchen—that are the same scale as the mirrored voids. Hence, it is occasionally difficult differentiating real space from illusionary space.

In a New York apartment approximately 18′ x 15′ with a 10′ high ceiling, eating, sleeping and conversation areas coexist. The furniture is pared down to its simplest, most abstract elements, and the space is given a dose of complexity with cutouts, mirrors and concealed, dramatic lighting, looking far larger than its actual dimensions (Fig. 3-3).

In a bathroom by El-Zoghby, it is especially difficult to detect the mirror, for the space is an exercise in illusionary and real voids. The bathtub is in a small room. The tub itself is reached via a circular cutout. That cutout is reflected by a mirrored cutout installed just over the tub. To complicate matters even more, the mirror over the tub reflects yet other mirrors in the space. It is doubtful if one would feel confined in such a complicated and seemingly endless space (Fig. 3-4).

In a somewhat simpler installation, a restaurant named Genghiz Khan's Bicycle, mirror occurs behind the bar. It reflects linear lighting elements, columns and circular window beyond (Fig. 3-5).

In a New York bedroom, mirror occurs behind the bed in a series of cutouts, and over the bed, defined by fake beams (Fig. 3-6).

Fortunately, El-Zoghby's mirrors usually have a low wall or ledge in front of them so that one doesn't walk through them (Fig.3-7).

Fig. 3-3. Gamal El-Zoghby: residence, New York, 1972. A series of voids, some real, some illusionary. Photo George Cserna.

Fig. 3-4. Gamal El-Zoghby: bathroom, New York, 1973. Only the shower head intrudes on the abstraction of this space. Photo Robert Perron.

Fig. 3-5. Gamal El-Zoghby: Genghiz Khan's Bicycle, New York, 1974. Photo Robert Perron.

Fig. 3-7. Gamal El-Zoghby: basement family room, New York, 1977. Photo Robert Perron.

Fig. 3-6. Gamal El-Zoghby: bedroom, New York, 1975. Photo Robert Perron.

Fig. 3-8. Robert A.M. Stern, Architects: Greenwich poolhouse, Greenwich, Conn., 1973–74. An elegantly mirrored void. Photo Ed Stoecklein.

In an elegantly mirrored corner in a poolhouse by Robert A.M. Stern, the mirror is carefully inserted in the wall rather than being applied to the wall. Also, since the mirrored opening is irregular, one doesn't expect to see mirror, for it is usually in more regular shapes (Fig. 3-8).

One of the most unusual recent mirrored installations was created by Shepley, Bulfinch, Richardson and Abbott at the Penobscot Bay Medical Center in Rockport, Maine.

A wealthy donor had specified in his will that he was leaving money for a colonial-style addition to the original hospital. As it turned out, a totally new hospital was built, not just a wing. In order to conform to the spirit of the will, however, it was decided to design one of the waiting rooms in the colonial style. The only problem was that the spaces provided for the waiting rooms were right triangles in the floor plan, hardly suitable to a traditional arrangement of furniture.

Enter illusionism. The architects decided to make the triangular room look like a square room. They lined the diagonal wall with mirror, then installed a colonial-style colonnade on the other two sides of the space. The mirror then completed the form, making the space look like it was four-sided instead of three-sided. And with assiduous attention to detail, the designers attached semicircular tables and semicircular lighting fixtures right next to the glass so that the mirror would complete their forms also (Fig. 3-9).

For these designers, the mirror was supposed to resemble pure, open space rather than a sheet of glass with a reflective backing. Of course, the joints, and the fact that one would eventually see one's reflection gave the illusionism a transitory nature. Indeed, it is hoped that groups of four did not attempt to sit at those little semicircular tables.

Moore, Grover, Harper used a similar idea at an office in New Jersey, although in a more complex and ambiguous fashion (Plate 7). They installed a mirror which visually completed a semicircular dome. The dome itself was comprised of fan-shaped sections, some open and some solid. The open sections admitted light from a skylight into the room below. The closed sections were themselves covered in a mirrored material, resulting in a light and shimmering fantasy. One is reminded of Sir John Soane's breakfast room which had a spherical dome, skylights above, and numerous mirrors. In Moore, Allen and Lyndon's book *The Place of Houses*, they wrote that Soane's breakfast room was "a particularly rich illustration of our suggestion to keep the myth up off the floor. The complex subdivisions overhead, made magic by the mirrors, along with the illusion of depth in the tiny pictures, contribute to this room, seeming at once miniscule and cosmic."

In the Moore, Grover, Harper design, the mirror that is placed flat on the wall is particularly illusionary since it is above eye level for the most part, and since the dome is an unexpected architectural feature, compelling attention (Fig. 3-10).

At their County Federal Savings and Loan, mirror is placed at the rear of the space. Once again, illusionism is likely to be transitory, since one can see one's reflection in the mirror. However, the sense of depth is particularly strong in the mirror due to the dynamic progression of architectural elements. In this case, the elements are angled walls with exposed fluorescent tubes attached to them (Fig. 3-11).

Victor Lundy, in his 1962 I. Miller Shoe Salon, also used mirror at the back of the space, duplicating the columns and billowing ceiling in the foreground. One has difficulty in determining which columns are real and which are illusionary. Also, since the mirror is usually juxtaposed with the natural wood materials, and the lighting level is low, the naturally greenish tint of the mirror is less obvious (Fig. 3-12).

At the time of Lundy's mirrored installation, such large-scale uses of mirror were unusual. Over a decade later, when Hellmuth, Obata and Kassabaum placed mirror at

Fig. 3-9. Shepley, Bulfinch, Richardson and Abbott: Penobscot Bay Medical Center, Rockport, Maine, 1975. The mirror makes a triangular room look like a square room. Photo Steve Rosenthal.

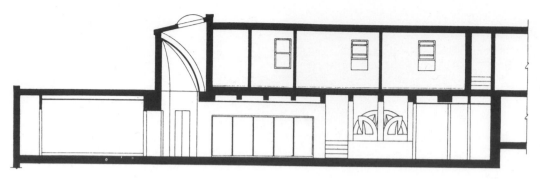

Fig. 3-10. Moore, Grover, Harper; Charles W. Moore, partner in charge of design and Mark Simon: Gund Investment Co. Office, Princeton, New Jersey, 1980. Mirror on the upper wall completed the form of the dome.

Fig. 3-11. Moore, Grover, Harper: County Federal Savings and Loan, Green Farms, Conn., 1973. An enchanced sense of depth by mirroring the progression of architectural elements. Photo Robert Perron.

the side of their Greenwich Savings Bank, mirror was less unusual in such installations. However, the HOK mirror had a different twist to it (Fig. 3-13, Plate 8).

From a pragmatic standpoint, the mirror was to help patrons understand that the banking level was one flight up, above the main entrance. But then, the designers heightened the illusionary depth in the mirror.

They cantilevered from the mirror a semicircular neon logo and cylindrical, silvery benches. The mirror had the dual effect of completing the form of those objects, and making them appear suspended in space. One *knows* that those are mirrors, yet there are those benches, looking as if they are about to float away. The illusionism is so strong that one hesitates to sit on those little spaceship facsimiles. Whereas in the past, mirrors were traditionally treated as a void surrounded by solids, the mirror at Greenwich Savings is most decidedly a void being penetrated by solids.

APPROXIMATIONS OF GLASS

The mirror in Richard Neutra's designs is particularly confusing, looking at times like clear glass, at other times like exterior space, at other times like interior space.

In the picture of his garden, the mirror occurs out-of-doors, adjacent to clear glass walls. Due to the complexity of the composition, the mirror is difficult to detect. According to Julius Shulman, when he and his assistant photographed Neutra's work, they had to walk around with hands outstretched so as not to shatter any illusionary space (Figs. 3-14 and 3-15).

Mirror is juxtaposed with clear glass in Neutra's living room in such a complicated way that the clerestory over the fireplace is almost impossible to decipher. Which piece of glass conveys illusions, which conveys views into another room (Fig. 3-16)? A bedroom with a mirrored wall seems all space (Fig. 3-17).

Fig. 3-12. Victor Lundy: I. Miller Shoe Salon, New York City, 1962. An early use of illusionism. Which columns are real? Photo George Cserna.

Fig. 3-14. Richard Neutra: own residence, Los Angeles, 1966. The mirror is placed in a panel out-of-doors. Not a void in a wall, but a plane. Photo Julius Shulman.

Fig. 3-13. Hellmuth, Obata and Kassabaum; Kahn and Jacobs; Vignelli Associates, graphics consultants: Greenwich Savings Bank branch, New York, 1975. A heightened sense of space due to the cantilevered logo and benches. Photo Barbara Martin.

Fig. 3-16. Richard Neutra: own residence, Los Angeles, 1963. Mirror and clear glass juxtaposed in a transom area adjacent to the fireplace. Photo Julius Shulman.

Fig. 3-15. Richard Neutra: own residence, Los Angeles, 1966. The mirror is difficult to detect, looking like another sheet of glass. Photo Julius Shulman.

Roche/Dinkeloo combined mirror with glass in conference rooms at Richardson-Merrell in Connecticut. From the conference table, one has the sense that one is looking into two other rooms and into the garden beyond through clear glass partitions (Fig. 3-18).

Jacobsen has also combined mirror with glass, and substituted mirror for glass. Bay windows often had a clear plexiglas front, top and bottom with mirror on the side supports (Fig. 3-19). In a Kentucky house, set upon an open meadow, Jacobsen created the expression of a pavilion with round brick piers and narrow glass inserts. The system of glass and column continued, even into the interior of the house. However, in locations requiring privacy, such as bathrooms, he inserted glass with a mirror finish on both sides. Since the panes of mirror were narrow and recessed, they did not easily convey the reflection of a passing person. They conveyed the illusion of clear glass further by reflecting the round columns, thus making them appear to be continuous.

In an earlier, pitched roof house, Jacobsen had installed a clear glass transom over a door, under the highest point of the roof. He noted that were he to execute the design again, he would have installed mirror over the door instead. It would have looked the same as the clear glass transom, and as an added benefit, it would have been cheaper since it would have allowed for the introduction of diagonal bracing behind its surface.

Perhaps the most deceptive mirror of all is the mirror one can look at directly and not see one's reflection. The infinity chamber is such an installation. It consists of basically a shallow box. The outer layer of the box is reflective glass or one-way mirror. The inner layer is

Fig. 3-17. Richard Neutra: own residence, Los Angeles, 1963. The mirror expands tiny dimensions of room and vistas beyond.

62

Fig. 3-18. Kevin Roche, John Dinkeloo and Associates: Richardson-Merrell Inc. Headquarters, Wilton, Conn., 1974. Rather than resembling pure space, this mirror resembles clear glass.

regular mirror. Sandwiched between the two are electric lights. The point of the lights is to make the interior of the box brighter than its exterior. Hence, the person standing outside the box sees through the one-way mirror and into a series of endless reflections on the interior of the box (Fig. 3-20). Such units have been fabricated into picture frame units on sale in department stores, and are available to architects in large-scale units that can be ganged together.

Paul Rudolph has experimented with infinity chambers. Working with a lighting manufacturer, he has designed various shapes and sizes with small custom lights, spaced in geometric configurations. Interestingly, the foyer of Rudolph's Manhattan office is filled with his varied mirrored designs: mirrored hanging screens, coffee tables, infinity chambers, even the ceiling in the elevator lobby (Plate 10).

When Rudolph was chairman of the architectural department at Yale, his work and philosophy were expressive of open and flowing architectural spaces. However, in the 1970's, one could see that his approach to space, like that of so many other architects, had developed an illusionary component. At times, the illusions were difficult to detect.

Fig. 3-19. Hugh Newell Jacobsen: Design House, Reston, Virginia, 1979. Plexiglas bay window and mirrored side supports. *(Courtesy Reston Development Corporation)*

Fig. 3-20. Wiard and Burwell: logo for Gannett Co., Rochester, New York, 1978. An infinity chamber, a combination of one-way mirror, lights, and regular mirror. Photo Ted Kawalerski.

4
Dynamic Effects

*But now the walls might disappear, the ceilings too,
and—yes—the floors as well. Why not? In certain cases.
Nicely calculated effects of this sort might amplify and
transform a cabinet into a realm, a room into bewildering
vistas and avenues, a single unit into unlimited areas of
color, pattern and form.*

Frank Lloyd Wright
Architectural Record, 1928

As we have seen in previous discussions, the reflected ray from the mirror's surface can be shifted or multiplied to creat a diversity of effects. The magician used the angled mirror in order to create the effect of complete transparency. Humphrey Repton in the nineteenth century used the same angled mirror as a means of shifting views from around the corner. Modern-day architects Charles Gwathmey and Susana Torre used the angled mirror for similar reasons.

Gwathmey considered the angled mirror installation at his Chicago Vidal Sassoon salon one of his most interesting uses of mirror (Fig. 4-1). By being angled, it conveyed views of "where you have been and where you are going," according to him. It conveyed a view of the shopping mall, the public space, through an adjacent glass window. It also conveyed the view of the salon itself, the private space. When in the shopping mall, one would be prevented from looking directly into the salon itself due to the angled mirror. So its function was dual: to convey views as well as to block views (Fig. 4-2).

Gwathmey used a similarly angled mirror in the entrance to the Evans Partnership. It gives one the impression that one is looking at multiple curved glass block walls, when in fact one is an illusionary image (Fig. 4-3).

Susana Torre's angled mirrors occur in a law office she designed in Manhattan. She placed mirrors at the outside corners of a U-shaped corridor at 45 degree angles. She insisted that the mirrors were not intended to be illusionary or decorative, but to convey views around the corners. She described her attitude:

At the risk of sounding pedantic, I must insist that my work not be presented in the context of a 'decorative' use of mirror (i.e., as mirrors are generally used, to 'enlarge' a small space, or to create amusing distortions, etc.). I use mirrors in an extremely precise way, to expand on, as well as to provide the meaning of my architecture . . . The two mirrors placed at 45 degree angles in the corners of the U-shaped corridor 'straighten' the corridor visually. The idea is not to disorient the observer but rather to make him/her realize what the circulation space actually is. This may seem paradoxical, but actually functions that way in the space itself.

One of Torre's mirrors is juxtaposed with a Duane Hansen sculpture of an exceedingly realistic looking rock star. The juxtaposition creates a startling dialogue between what is

real and what is unreal. Even today, such an installation is unusual in an office setting (Figs. 4-4 and 4-5).

If one analyzes Torre's installation in plan, one will see that she created what is in effect a reverse periscope. The mirrors are both angled, but not parallel. As the light is reflected from one mirror to another, the direction of the final emergent ray is the reverse of the incident ray. Hence, images conveyed by that emergent ray are reversed (Fig. 4-6).

In a true periscope, the effect is different. Since both mirrors are parallel, the final emergent ray is traveling in the same direction as the incident ray. Hence, the image is not reversed. In a submarine, a submariner looking through a periscope will see other ships moving in the correct direction rather than reversed, because of the direction of the emergent ray (Fig. 4-7).

The traditional maritime periscope shifts views from above the water down below the water. Gunnar Birkerts took the concept of the periscope and installed it on the facade of his Museum of Glass in Corning, New York. He wanted to shield the glass wall of the building from direct solar radiation, yet he did not want to block views, frequently a negative side effect of solid sun control devices. So he shielded the glass with an angled mirrored "eyebrow" above. The mirror on the sunshade received images from a second mirror, placed on the angled spandrel below. As the observer looked out of the window he or she would see two views of the out-of-doors simultaneously, one real, the other illusionary. According to Birkerts, this is the first time the concept of the periscope has been used in an architectural fashion such as this (Fig. 4-8).

Fig. 4-1. Gwathmey-Siegel: Vidal Sassoon salon, Chicago, 1976. The angled mirror used to shift views.

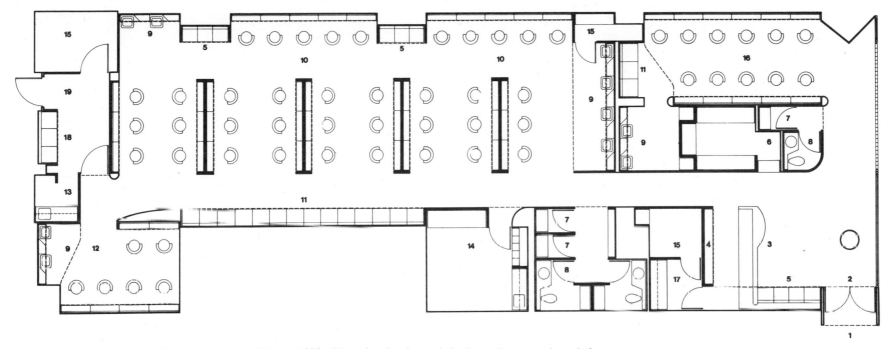

Fig. 4-2. Gwathmey-Siegel: Vidal Sassoon salon, Chicago, 1976. Floor plan showing angled mirror adjacent to glass window.

FLOOR PLAN

1 MALL
2 ENTRY
3 RECEPTION
4 DISPLAY
5 WAITING
6 COATS
7 DRESSING
8 TOILETS
9 WASHING
10 CUTTING (WOMEN)
11 DRYING
12 TRICOLOGY
13 DISPENSARY
14 STAFF ROOM
15 STORAGE
16 CUTTING (MEN)
17 MANAGERS OFFICE
18 LAUNDRY
19 SERVICE

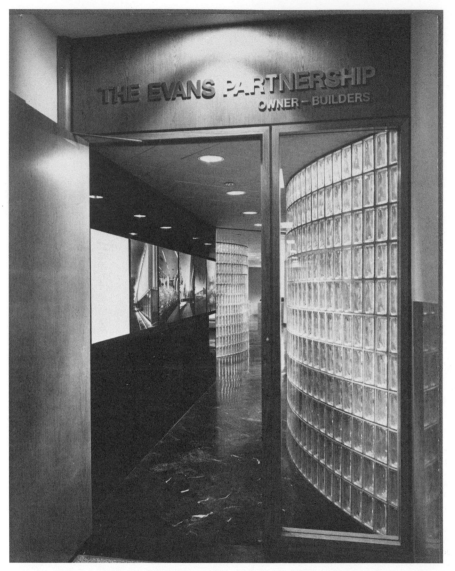

Fig. 4-3. Gwathmey-Siegel: Evans Partnership, New York, 1977. An angled mirror in the distance. Photo Otto Baitz.

Fig. 4-4. Susana Torre: law office, New York, 1976. An angled mirror that conveys views down the corridor, and views of the realistic looking sculpture. Photo Norman McGrath.

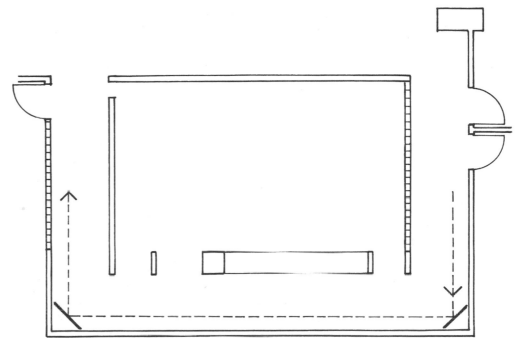

Fig. 4-6. Susana Torre: law office, New York, 1976. Floor plan. Torre has in plan created a reverse periscope. The emergent ray travels in the opposite direction as the incident ray.

Fig. 4-5. Susana Torre: law office, New York, 1976. Photo Norman McGrath.

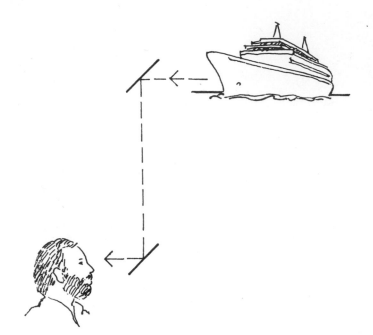

Fig. 4-7. Sketch of periscope. The emergent ray travels in the same direction as the incident ray. Hence, images are not reversed. Ships, for instance, will appear to move in the correct direction.

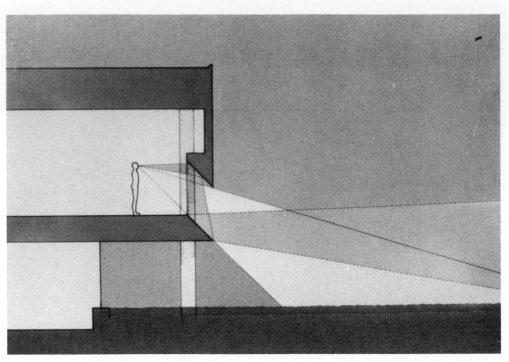

Fig. 4-8. Gunnar Birkerts and Associates: Museum of Glass, Corning Glass Works, Corning, New York, 1980. A periscope used as a sunshade.

One aspect of the Corning installation that could prove disconcerting was that the illusionary image would be seen above the real image. However, the illusionary image would have originated below the real image. Hence, one could look out the window and see sky, for instance, then rooftops hovering above that. In a sense, it was difficult determining if this was an "interior" use of the mirror or an "exterior" use. It's really both. The mirrors occur outside the building, but are intended to be seen inside the building.

One can imagine numerous applications of angled mirrors, either solitary or combined. This writer spent most of the time typing this book while gazing out the window. However, it was not a real window but the illusion of the window as conveyed by an angled mirror. Now that many of us are city dwellers with often inadequate views, the angled mirror can help to compensate for insufficient windows, or windows on only one side of a room.

Some architects have begun to experiment with buildings underground, or partially underground, as a means of conserving energy since the earth is an excellent natural insulator. Obviously, the lack of views and windows could prove depressing.

One could take the periscope effect, as used by Birkerts, and apply it to an underground dwelling (Fig. 4-9). An angled illusionary window could occur below ground, reflecting images from another mirror above ground. One thing to note, however, is that there will be a natural diminution of view as conveyed from one mirror to another. In true periscopes, lenses are installed between the mirrors to conpensate for this effect. The architect experimenting with this technique would do well to make models first.

As for models, periscopes have distinct utility there. In the offices of Saarinen and Roche/Dinkeloo, the architects built small periscopes in order to understand better the small scale of their architectural models. The periscopes helped them get "into" the models rather than merely looking at them from above.

Fig. 4-9. A mirror glass awning window, installed on the inside, can project a view to a lower mirror. The awning also reduces glare in the day, and heat loss at night when closed.

If one could somehow stand between the two mirrors forming a periscope, one would notice the effect of *glaces à répétitions* created by the parallel faces. Sometimes—but not always—this is a highly decorative mirror.

As noted earlier, Peter Blake used face-to-face mirrors at the exhibition, "America Builds" in order to demonstrate the sense of verticality or horizontality in American steel and glass buildings. It is fascinating to see how the observers are standing in the photograph, somewhat hesitantly, as if afraid to fall off the precipice. Blake said, "They were getting woozy." Blake's installation is intended less to be decorative than to convey information about a type of architecture. The vertical imagery is particularly potent (Fig. 4-10).

In Derek Romley's 1968 apartment, mirrors placed on movable storage cabinets are intended more to disorient than to decorate. When the cabinets are grouped around each other, the mirrors create multiple reflections, conveying the impression of infinity. Yet, because of their free-standing forms, the cabinets make the mirrors clearly finite (Fig. 4-11).

Some architects who use a lot of mirrors do not care for the effect of multiple reflections. Gwathmey, for instance, rarely uses it, saying he does not care for the "tunnel" effect. And there are times when combined mirrors seem too much of a good thing.

In New York, there is a showroom that specializes in expensive mirrored furnishings and installations. Everything in the showroom is mirrored. Mirrored walls reflect

Fig. 4-10. Peter Blake: exhibition "America Builds," West Berlin, 1957. A rare use of mirror in the 1950's. *Glaces à répétitions* intended to inform about the verticality of the glass skyscraper. Photo Gnilka.

other mirrored walls which reflect mirrored chests, tables and bed platforms. The effect is cold, murky and depressing. This is mainly due to fact that most mirrors are made of glass that has a slightly greenish tint to it. When the views from mirror to mirror are multiplied, the greenish effect is intensified. This darkening effect is particularly noticeable when tinted glasses are combined. One sees an even greater effect of darkness in the distance.

For this reason, the French invariably used as clear a glass as they could obtain and combined it with chandeliers. Hence, their *glaces à répétitions* were truly decorative as one saw an expansion of tiny pinpoints of light in the distance.

Paul Rudolph's 1976 apartment features bundles of tiny electric lights suspended just in front of the mirrors at opposite sides of the room (Figs. 4-12 and 4-13). The total effect is celestial. Rudolph said in *House and Garden*:

Pinpoints of light can be used in a room to break down the scale and make the space seem more human. Combined with mirror they will expand the space both vertically and horizontally. And, the wall that glows is so much more interesting than the standard lighting fixture.

The men's room at La Folie restaurant in New York cheerfully celebrates colored neon lights in its multiple reflections (Fig. 4-14). The multiple mirrored guest bathroom designed by Short and Ford seems far larger than its actual dimensions (Fig. 4-15).

Fig. 4-11. Derek Romley: own apartment, New York, 1968. *Glaces à répétitions* intended neither to inform nor to decorate, but to disorient. Photo Louis Reens.

73

Fig. 4-13. Paul Rudolph: own apartment, New York, 1976. Photo Donald Luckenbill.

Fig. 4-12. Paul Rudolph: own apartment, New York, 1976. A frankly decorative installation of *glaces à répétitions* and small-scale electric lights. Photo Donald Luckenbill.

Fig. 4-14. Fred Victoria and David Barrett: men's room at La Folie restaurant, New York, 1977. A jazzy use of mirrors and neon. Photo Robert Perron.

Fig. 4-15. Short and Ford: Guernsey Hall Condominium, Princeton, New Jersey, 1973. A small space amplified with multiple mirrors. Photo Otto Baitz.

75

Perhaps the most dazzling, intriguing and memorable of the recent installations of *glaces à répétitions* was created by Roche/Dinkeloo at the U.N. Plaza restaurant in New York (Fig. 4-16, Plate 3).

The space was windowless and interior, yet Roche wanted it to have the "backyard" quality the one might find in a secluded European outdoor restaurant. The architects created an effect of elegant spatial complexity with remarkably simple means. In a scooped-out area of the ceiling, they placed canted panels of stretched, mirrored plastic. Below the mirrors, they installed a delicate trellis, randomly lighted. The resultant reflections are so complex that people walk in from the street merely to look at the ceiling, trying to figure it out.

Roche/Dinkeloo created a similarly complex effect in the elevator lobby atop the Worcester County National Bank Building. It consisted of mirrors covering the elevator core, mirrored beams on the sloping ceiling, and mirror on the angled top of another elevator core. In the recent past, the elevator cores would have been clad in travertine or some similarly substantial material, creating an unambiguous statement. But in this curtain wall building, one has difficulty in determining where the real view ends and the illusionary view commences (Figs. 4-17 and 4-18).

Fig. 4-16. Kevin Roche, John Dinkeloo and Associates: U.N. Plaza restaurant, New York, 1975. One of the most dazzling recent examples of *glaces à répétitions*.

Fig. 4-17. Kevin Roche, John Dinkeloo and Associates: Worcester County National Bank Building, Worcester, Mass., 1974. A mirrored elevator lobby.

Fig. 4-18. Kevin Roche, John Dinkeloo and Associates: Worcester County National Bank Building, Worcester, Mass., 1974.

Fig. 4-19. Kevin Roche, John Dinkeloo and Associates: Worcester County National Bank Building, Worcester, Mass., 1974. What looks like glass pavilions is actually the service core covered with mirror.

As one enters the dining room just beyond the elevator lobby, one sees what appear to be dining spaces which are interlocking and enclosed by glass walls. What one is really seeing are the service cores, mirrored so as to become less obtrusive and to amplify the breathtaking view of the Connecticut River Valley below. At the top of the partitions were installed lighting fixtures, resulting in indirect light intensified due to reflection from the mirror (Fig. 4-19).

The effect of *glaces à répétitions* has been used by designers on the exterior of mirror glass buildings in order to make them more visually interesting. But one gets a glimpse of that approach in the photograph of the interior of Johnson/Burgee's IDS Center in Minneapolis. The fragmented mirror glass wall of the main building reflects its own walls, and the skylight beyond (Fig. 4-20).

For Washington, D.C. artist Rockne Krebs, mirrors are a dynamic tool, not for the images on the mirrors, but for beams of light that he has reflected from mirrors. Working at an architectural scale, he has created light sculptures using mirrors and sunlight or mirrors and laser beams. At the Philip M. Stern house, a laser beam emerges from an upper-story window, and is redirected by mirrors installed on trees. Thus, with relatively simple means, a complex canopy of light is created (Fig. 4-21).

Obviously, many effects are possible through the shifting or multiplication of light rays. Though the technique might be constant, the appearance often won't be.

Fig. 4-20. Philip Johnson and John Burgee: IDS Center, Minneapolis, Minn., 1973. A large-scale version of *glaces à répétitions*. Photo Alexandre Georges.

Fig. 4-21. Rockne Krebs, artist: laser beam sculpture at Philip M. Stern house, Washington, D.C., 1970. Laser beam is redirected with mirrors.

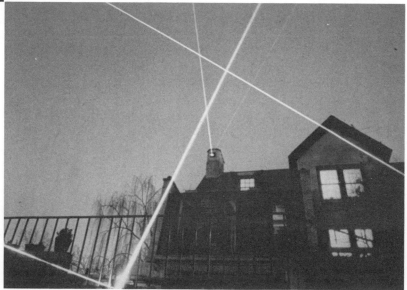

5
The Exterior Mirror

The mirror is seen in nature in the surfaces of lakes
in the hollows of the mountains and in the pools deep in
the shadow of the trees; in winding ribbons of the rivers that
catch and give back the flying birds, clouds and blue sky.
A dreary thing to have that element leave the landscape.
It may be as refreshing and beautifying in architecture, if
architecturally used. To use it so is not easy, for the
tendency toward the tawdry is present in any use of the mirror.

Frank Lloyd Wright
Architectural Record, 1928

At the time of Wright's quote, mirrors were popular interior decorative items. The mirror glass building had not been developed, and architects did not experiment with solar mirrors. In those days, if an architect used a mirror out-of-doors, it might occur in a genteel little garden niche as a way of doubling certain views, and expanding the size of the garden.

When Wright wrote of the mirror in nature, he was writing rhetorically. He had no way of foreseeing the ubiquitousness and enormous scale of mirror glass buildings that were to occur. Nor could he envision the future experiments with solar mirrors. These exterior mirrors were to appear on the architectural scene shortly after his death in 1959.

Mirror glass buildings were typically curtain wall office buildings clad in glass with a reflective coating. That coating would reflect away heat-producing sunlight (Fig. 5-1). In essence, that glass was the one-way mirror, refined to withstand the rigors of exterior use. Since light was reflected from the glass, the buildings would look like giant mirrors when the sun was shining. On their interiors, where they were darker, they would appear transparent to the occupants. At night, when the lighting conditions were reversed, the buildings would look transparent from the outside and reflective from the inside.

These buildings were a direct outgrowth of the international style curtain wall building. They were developed because of the problem of heat buildup in the all-glass curtain wall building. It is ironical that the most visible and prevalent use of the mirror, the mirror on an entire multistory building, was to emerge from the building type that had made the interior mirror virtually obsolete.

Wright noted how difficult it was to design with mirrors. This was certainly true with the enormous scale of mirror glass buildings. Since they looked opaque from the outside, they often looked blank and scaleless. Since they reflected images on their surfaces while they were keeping interiors cooler, their images were often ill-considered. Reflected glare, and along with it heat, were also problems.

Though these mirrors covering entire buildings out-of-doors were not vastly different from mirrors covering walls on the interior, the effects were quite different. As we have seen, the interior mirror could, with careful handling, be so unobtrusive that it would cease to exist on a perceptual

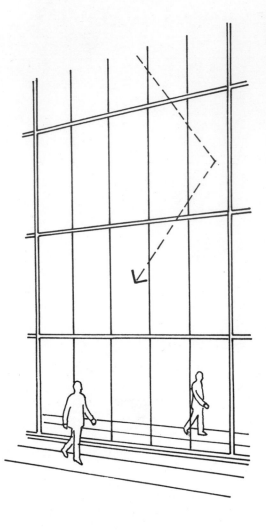

Fig. 5-1. One-way mirrors are used as an exterior cladding. They prevent a certain percentage of sunlight from entering the building.

basis. It could resemble pure space. But somehow, these exterior mirrors never managed that feat. They always remained visible. Even with minimal mullion systems, their joints could never disappear. The glass often had visual imperfections resulting in distorted images. Their *raison d'être*, the sun, would invariably glance off, making them bright and noticeable. Instead of disappearing, these buildings were very often the brightest thing around.

The mirrored buildings that succeeded aesthetically did so because architects realized this important limitation. They realized that the surfaces wouldn't just fade away into a mirage of clouds and sky, but would have a curiously visible presence. They would remain forever mirrored surfaces, at the forefront of our consciousness. They understood that to give these buildings architectural identity, they would have to manipulate those surfaces with care. They understood that the occasionally beautiful reflections couldn't do all the work.

Mirror glass buildings usually treated solar energy as a negative. It was something to be reflected away. The sun's heat was something that simply wasn't wanted in building interiors. Solar mirrors, on the other hand, treated solar energy as a positive. They redirected the sun's energy, often concentrating it, toward some form of receiver (Fig. 5-2). Depending on design and number of mirrors, temperatures in solar installations have ranged from 200 to 7000 degrees Fahrenheit. That energy has been used for a variety of purposes. It might be something as simple as melting snow on a doorstep, to something as complex as fusing metals in a high-temperature installation.

Solar mirrors are one approach to solar energy. As solar energy is still experimental, so also are solar mirrors. They are nowhere nearly as prevalent as mirror glass buildings. And rather than covering an entire building, they usually perform a supplementary role, near or on one portion of the building.

Many of these solar mirrors are beautiful, often without attempting to be beautiful. Sometimes the forms are as

PLATE 1. Kevin Roche, John Dinkeloo and Associates:
Richardson-Merrell, Inc. Headquarters, Wilton, Connecticut, 1974.

PLATE 2. Kevin Roche, John Dinkeloo and Associates:
John Deere Addition, Moline, Illinois, 1978.

PLATE 3. Kevin Roche, John Dinkeloo and Associates: United Nations Plaza restaurant, New York, 1975.

PLATE 5. Gwathmey-Siegel: Shezan restaurant, New York, 1976. Photo Norman McGrath.

PLATE 4. Gwathmey-Siegel: Shezan restaurant, New York, 1976. Photo Norman McGrath.

PLATE 7. Moore, Grover, Harper, Charles W. Moore, partner in charge of design and Mark Simon: Gund Investment Co. Office, Princeton, New Jersey, 1980. Photo Norman McGrath.

PLATE 6. Warren Platner Associates: Providence Journal Headquarters, Providence, Rhode Island, 1976. Photo Erza Stoller.

PLATE 8. Hellmuth, Obata and Kassabaum; Kahn and Jacobs; Vignelli Associates, graphics consultants: Greenwich Savings Bank branch, New York, 1975. Photo Barbara Martin.

PLATE 11. Bromley/Jacobsen: Xavier's apartment, New York, 1978. Photo Jaime Ardiles-Arce.

PLATE 9. Robert A.M. Stern: residence, Westchester County, New York, 1976. Photo Ed Stoecklein.

PLATE 10. Paul Rudolph: Infinity Chambers, New York, 1978. Photo Robert Perron.

PLATE 12. Eero Saarinen and Associates: Bell Labs, Holmdel, New Jersey, 1962–1967. (*Courtesy Bell Labs.*)

PLATE 13. I. M. Pei and Partners, architects, Henry N. Cobb design partner: John Hancock Tower, Boston, Massachusetts, 1976. Photo George Cserna.

PLATE 14. Arthur Erickson: Canadian Pavilion, Expo '70, Osaka, Japan, 1970.

PLATE 15. Cesar Pelli for Daniel Mann, Johnson and Mendenhall:
California Jewelry Mart, Los Angeles, 1968. Photo Arthur Golding.

PLATE 16. Welton Becket Associates: Reunion Center, Dallas, Texas, 1978.
Photo Balthazar Korab.

PLATE 17. Welton Becket Associates: Reunion Center, Dallas, Texas, 1978.
Photo Balthazar Korab.

PLATE 18. Welton Becket Associates: Reunion Center, Dallas, Texas, 1978.
Photo Balthazar Korab.

PLATE 19. Paolo Riani Associates: Hampden Country Club, Springfield, Massachusetts, 1974.

PLATE 20. Paolo Riani Associates: Hampden Country Club, Springfield, Massachusetts, 1974.

PLATE 21. Caudill Rowlett Scott, Inc.; Shropshire, Boots, Reid and Associates: Indiana Bell Telephone Switching Center, Columbus, Indiana, 1978. Photo Balthazar Korab.

PLATE 22. Caudill Rowlett Scott, Inc.; Shropshire, Boots,
Reid and Associates: Indiana Bell Telephone
Switching Center, Columbus, Indiana, 1978.
Photo Balthazar Korab.

PLATE 23. Cesar Pelli for Gruen Associates: United States Embassy, Tokyo, 1972.

PLATE 24. Hugh Stubbins and Associates: Citicorp Center, New York, 1979.
Photo Norman McGrath.

PLATE 24. Hugh Stubbins and Associates: Citicorp Center, New York, 1979.
Photo Norman McGrath.

PLATE 25. The Architects' Collaborative Inc.: Johns-Manville World Headquarters Building, near Denver, 1976. Photo Nick Wheeler.

PLATE 26. The Architects' Collaborative Inc.: Johns-Manville World Headquarters Building, near Denver, 1976. Photo Nick Wheeler.

PLATE 27. Solar furnace at Odeillo, France. Photo Robert Perron.

PLATE 28. Michael Jantzen: vacation house, Carlyle, Illinois, 1978.

PLATE 29. Michael Jantzen: vacation house, Carlyle, Illinois, 1978.

PLATE 30. James Lambeth: Yocum Lodge, Snowmass at Aspen,
Colorado, 1973.

PLATE 31. James Lambeth: Strawberry Fields Apartments, Springfield, Missouri, 1976.

elegant and simple as the grain silos and steamships that Le Corbusier loved. Sometimes the reflections, a faceted image of sky and clouds, are pleasant to observe. And there is still an element of surprise about these highly technical mirrors. It is rather like seeing an acquaintance, whom we always suspected had a rather frivolous nature, engaged in a new and serious pursuit. Can we really take him seriously?

The mirror today is a more refined material than first appeared at Versailles. It often reflects light more perfectly than before. It's just that today, light is more often than not a direct ray cast by the sun. For architects, the mirror is no longer basically an interior material.

BACKGROUND

For early man, mirrors were small hand held objects generally made of highly polished metal. As we have seen, these early mirrors had no architectural applicability. They were generally used as an aid to personal adornment. However, they had a place outside the boudoir as well. Because of their unique ability to reflect light in a predictable fashion, they were useful in creating fire. Mirrors could focus sunlight toward a clump of leaves or twigs, and instantly produce this life-sustaining force.

Convex pieces of glass, called burning glasses, were used in a similar fashion. Burning glasses have been found in the ruins of ancient Nineveh, and date from the seventh century BC. And there exists the legend that Archimedes, the Greek mathematician, focused an array of mirrors on an approaching Roman fleet in 214 BC and set it ablaze with the concentration of sunlight.

During the Renaissance, Leonardo da Vinci explored many aspects of the mirror. Not only did he write notes backward with the aid of a mirror, but he also observed that art works could be admirably judged if reflected in a mirror, tigers could be foiled by a mirror if a hunter were taking a cub, face-to-face mirrors would reflect candles in an infinite progression, and that the cold surface of the

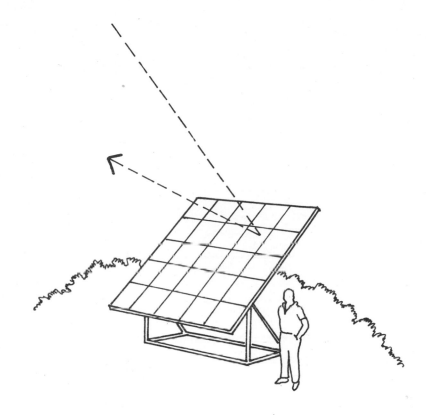

Fig. 5-2. Solar mirrors redirect sunlight toward a focal point.

83

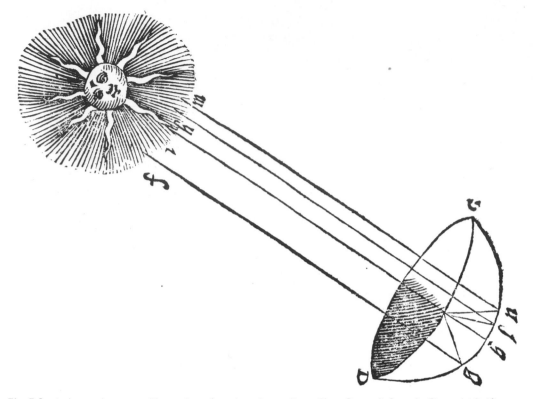

Fig. 5-3. A sixteenth centruy illustration of a solar mirror. From Fine, *Opere di Orontio Fineo del Delfimato* (*Courtesy Library of Congress*)

mirror could also produce heat. He performed a number of experiments with concave mirrors, sometimes using fire as the energy source, sometimes the sun itself. He likened the mirror to a bellows. Where one could produce an extreme amount of cold, the other could produce an extreme amount of heat.

As scientific experimentation increased, solar mirrors were often useful tools (Fig. 5-3). At the beginning of the seventeenth century, Athanasius Kircher, a German writer, described what is considered by some to be the first solar furnace. He used a concave mirror to focus the sun's rays toward a container of water. The water would be vaporized, then condensed, resulting in distilled water. In the late eighteenth century, Joseph Priestley, the English chemist, used a burning glass to extract a new gas from mercuric oxide. The gas was six times purer than ordinary air. Antoine Lavoisier, the French chemist, repeated Priestley's experiments using two lenses. He named the new gas oxygen from the Greek word *oxys* for acid, because of its acid-forming properties. The scientists used energy from the sun rather than from other means because they wished to establish laws of chemical change without contamination.

On rare occasions, architects used mirrors out-of-doors. Mirrored garden ornaments appeared at Catherine the Great's palace of Potemkin, and designers sometimes inserted mirrors in garden walls as a means of expanding certain views. Nevertheless, these installations were peripheral.

Throughout the nineteenth century, scientists and engineers had the greatest interest in the mirror out-of-doors. Numerous solar steam engines were constructed. In the late 1870's, Abel Pifre, a French engineer, built a solar steam engine with concave mirrors to run a printing press. The product of this press was a newspaper appropriately entitled *Le Journal du Soleil*. The engine was exhibited at the World Exposition in 1878 in Paris and was a popular attraction at the show (Fig. 5-4).

Solar engines as well as solar irrigation projects continued to be tested into the early days of the twentieth century. But the ventures usually ended in failure. Though energy from the sun was free and abundant, it was diffuse. The means to concentrate all that free energy was expensive. Reflecting surfaces were initially costly and had to be maintained; if they were made of metal, they were often damaged by corrosive elements in the atmosphere. An early pioneer in solar reflectors, Carl Guntner, said in 1906, "I have arrived at the conclusion that the sun's rays may only be made serviceable for industrial purposes when it is possible to concentrate them at low cost and in immense quantities." His prediction was accurate, the mirror having little applicability for industrialized areas for more than half a century. It proved to be useful occasionally in primitive areas, however. During World War II, small solar still "survival kits" were designed by Dr. Maria Telkes and used by the U.S. Navy. Also, solar cookers, large umbrellalike devices, were developed for use in underdeveloped areas. Distributed by the United Nations in India, they could develop temperatures of up to 350 degrees Fahrenheit. Not terribly efficient, they would take a half hour to boil water.

As for the one-way mirror, it was an interior mirror in its early days. Like the early solar mirror, it had nothing to do with aesthetics. Initially, it was used in the scientific laboratory. Known in scientific parlance as the partially silvered or semitransparent mirror, it was a useful tool in a variety of precise measurements, such as determining the speed of light.

A French scientist named Armand Fizeau was the first to use mirrors in the measurement of light. Basically, the light was directed through a semitransparent mirror, sent a distance of over five miles, then reflected from another mirror back to its original source. Between the mirrors was a rotating disc which, when rotated rapidly enough, would block the return of the light ray. From the speed of the disc, the speed of light could be calculated. The method, with variations, persisted into the twentieth century when Michelson and Morley determined the speed of light to be 186,000 miles per second. With the speed of light determined, Albert Einstein could develop his theory of relativity, which was based on the unchanging speed of light.

Fig. 5-4. A nineteenth century printing press powered by the sun. From Pope, *Solar Heat, Its Practical Applications.* (*Courtesy Library of Congress*)

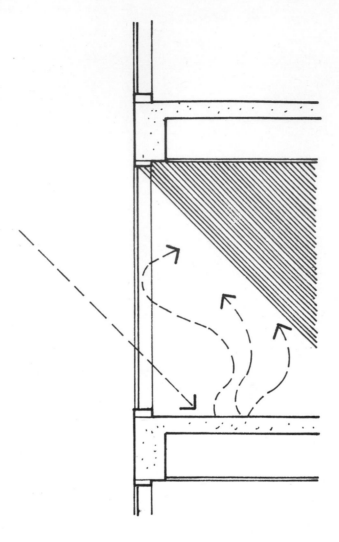

Fig. 5-5. The greenhouse effect. Solar light is transformed to infrared radiation, then trapped by the glass.

The semitransparent, or one-way mirror, also had a playful side. We have seen how it was a popular device in fairs and expositions in the nineteenth and early twentieth centuries. People undoubtedly made their own semitransparent mirrors at home, for *Scientific American,* which in the nineteenth century was a journal of recent inventions, gave precise "how-to" instructions.

In the twentieth century, the one-way mirror was a common tool in hospital observation wards. A patient would typically be in a brightly lighted space separated from the observer by a one-way mirror. Since the observer was in a darker space, he would be undetected by the patient.

In the early 1960's, the one-way mirror was used out-of-doors in a large-scale, architectural fashion. Suddenly, for architects, the mirror was no longer an interior material.

Beginnings of the Mirror Glass Building

In the distance shines our tomorrow. Hurray, three times hurray for our kingdom without force! Hurray for the transparent, the clear! Hurray for purity! Hurray for crystal! Hurray and again hurray for the fluid, the graceful, the angular, the sparkling, the flashing, the light—hurray for everlasting architecture!

Bruno Taut

Early modern theoreticians and designers viewed glass as more than just a convenient form of exterior cladding of buildings. It was also a symbol of a new society, free from the vicissitudes of war and poverty, living in clean and wholesome environments. The dark, ponderous buildings of the past would be replaced by towering prisms of glass.

Yet, when those visions became reality during the 1950's in the United States, they possessed a number of technical as well as aesthetic problems. One of the most severe of the technical problems was the intense heat that would build up in the spaces enclosed by those sheer glass walls.

In buildings of traditional masonry design, heat buildup had not been much of a problem since much of the solar radiation had been absorbed in the solid masonry walls. But in all glass buildings, something called "the greenhouse effect" occurred in what were designed not to be greenhouses at all but homes, offices and schools for people.

Curtain wall buildings would become intolerable for occupants on the south and west sides in particular, due to the direct solar radiation striking the glass. Solar radiation consists of differing wavelengths of energy, ranging from the shorter, more energetic ultraviolet rays, through visible light, then into the longer, less energetic infrared rays, X rays and radio waves. Most of the sun's heat is contained in the longer infrared rays which do not penetrate the glass. Glass is opaque to infrared radiation whereas it is transparent to visible light.

However, a curious thing would happen to the visible light that entered the building. Upon striking various objects, its wavelength would be transformed. It changed to infrared radiation. Since the glass was opaque to infrared radiation, that heat would remain in the interior space. This is the greenhouse effect (Fig. 5-5).

Designers agreed that the most effective means of dealing with direct solar radiation was the use of exterior shading devices. The sunlight would be blocked before it ever reached the glass. However, those devices were often expensive, subject to weathering, and often they would block views. The next best solution was the internal shading device such as curtains, window shades or venetian blinds. If made of a light, reflective material they would reflect visible light back through the glass before it had a chance to become transformed into infrared radiation (Fig. 5-6).

Another solution, particularly prevalent in academic institutions with new glass buildings which were rarely

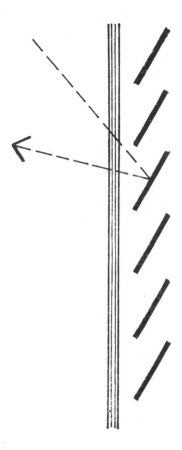

Fig. 5-6. Venetian blinds reflect sunlight back through the glass before it is transformed into infrared radiation.

Fig. 5-7. Skidmore, Owings and Merrill: Lever House, New York, 1952. Reflections on clear glass were considered a flaw in the 1950's.

air conditioned, was to tape aluminum foil or white paper to the glass. It was almost a plague that hit even the very best buildings of the most prominent architects. The shreds of paper and aluminum foil looked like pathetic distress signals.

In the 1950's, tinted glasses were developed to absorb a lot of the visible light that struck them. However, they were not ideal, since much of the energy that was absorbed by the glass would be radiated back into the interior of the building eventually.

Apart from the technical problem of heat buildup, there were also aesthetic difficulties with curtain wall buildings. Though Mies van der Rohe in 1919 had admitted that reflection was more important than shadow in the look of an all-glass building, particularly a tower, the ideal for such buildings was transparency. Yet, transparency was sometimes difficult to achieve, for ordinary window glass looks reflective when viewed obliquely, particularly if it is in shadow. Sometimes, ordinary glass performs similar to a mirror; though most light is transmitted through glass, some is reflected (Fig. 5-7).

In the Manufacturer's Trust Building by Skidmore, Owings, and Merrill in 1954, true transparency was achieved only by installing a brilliantly lighted suspended ceiling which counteracted the exterior reflections. *Architectural Forum* said of the bank in 1954:

When you look at most glass walled buildings in daylight you cannot see inside; instead, you see your ghost reflected darkly in the glass. But thanks to its bright interior, this is not true in the new bank. Because it is not true, this may be the first big building to truly fulfill architects' immaculate drafting board idea of glass as an invisible material.

True transparency was difficult to achieve. When it was achieved, architects often did not like the chaos that was revealed behind the elegant facades. Cluttered desks, crooked venetian blinds, water-stained draperies, not to mention draperies of different color, did not improve the appearance of a building. When Mies van der Rohe's 845 and 860 Lake Shore Drive apartments were completed, not only did they have glass outer walls, they also had outer draperies installed in all of the apartments so as to maintain a uniform appearance from the outside.

Another difficulty with curtain wall buildings was regularity. One of the foundations of the international style, it nonetheless made for a lot of buildings with a clear structural system boringly revealed. And when the system was so visible, it was difficult to break that system.

Yet, the curtain wall building was cheap to build. Since the exterior bearing wall had been eliminated and replaced by slender steel columns and beams, the foundations were smaller. The whole ensemble could be erected rapidly. Once the outer skin was perfected, it was virtually leakproof.

Given its obvious economies, the curtain wall building was a useful form in the 1950's. After 15 years of depression and war, there was an enormous need for new housing, hospitals, offices and schools in the United States. And that type of building satisfied much of that need quite efficiently.

When Eero Saarinen and Associates were working on the General Motors Technical Institute at the start of the 1950's, they realized that in the future, buildings would contain a considerable amount of glass. They were aware of the various technical problems. At this time, sealants were inadequate. Hence, panels would pop out of frames. As a result, they helped to develop the zipper gasket.

They noted the problem of heat buildup in all glass areas, and were dissatisfied with tinted glass and venetian blinds, the usual means of controlling the problem. As for venetian blinds, Saarinen thought it illogical to create expansive glass areas, only to cover them up with shading devices.

In the late 1950's, John Dinkeloo, then a partner in the firm, began to investigate the possibility of developing a

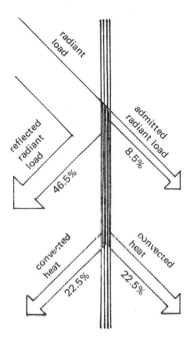

Fig. 5-8. Glass developed for Bell Labs by Kinney Glass.

Fig. 5-9. Eero Saarinen and Associates: Bell Laboratories, Holmdel, New Jersey, 1962. Initially, the building was clad in tinted glass except for one wall, not shown, of mirror glass. (*Courtesy Bell Labs*)

glass that could reflect light, and therefore heat, away from the building before it ever entered the wall. He thought that the one-way mirror could be developed for architectural use on the outside of buildings.

The one-way mirror had been commonly used in hospital observation wards. It consisted of a single sheet of glass with a thin silver coating on the back. Due to the thinness of the coating, some light would be transmitted through the glass, but a good portion of the light would be reflected away from the glass's surface. The main problem with such mirrors was that the silver film was unprotected and could be easily scratched. Dinkeloo sought a more durable system.

Most architects tend to work within the constraints that certain industrial products might impose on them. But Saarinen's firm attempted to work with industry to create new products where needed. Hence, it wasn't surprising that Dinkeloo went to small mirror and glass companies, namely Liberty Mirror in Philadelphia and Kinney Glass in Detroit, to create an improved one-way mirror.

One system they experimented with was the application of a reflective mylar onto a sheet of ordinary window glass. But it was difficult to achieve a uniform surface. The mylar frequently would buckle. The second system, and ultimately successful method, was developed by Kinney Glass. It consisted of the electrolytic deposition of metal on glass in a vacuum chamber. In order to protect the metal film, a second piece of glass was applied to the back.

Though the main impetus of the architects was to avoid heat buildup in all-glass buildings, they were quite aware of the aesthetic component from an early point. According to Kevin Roche, while the firm was working on the General Motors Technical Institute, another project came into the office. It was a brief project for Alison Jeffries which was never constructed. The design featured reflective aluminum spandrel panels. According to Roche, "We thought it would be nice if the glass matched the panels . . ." Obviously, in the best architectural firms, aesthetic and technical motivations are intertwined in a complex fashion.

The first building to utilize the new glass was Bell Laboratories in Holmdel, New Jersey, a Saarinen building constructed in 1962, a decade after the General Motors Technical Institute.

The building's function was to develop components for space satellites. Since the laboratories themselves required flexibility and precise temperature control, they were separated from the exterior wall by a perimeter corridor.

Initially, the building had mirror glass only on the south facade. Most of the building was clad in tinted glass. The reason was that the architects had been ahead of their time in specifying the new glass. Kinney was able to produce only enough for one wall (Fig. 5-8).

When the building doubled in size in five years, the client decided to replace all of the tinted glass with the new reflective glass. Nearly two acres of the original glass were replaced with the new product (Figs. 5-9 and 5-10, Plate 12).

The client was delighted with the heat-reducing characteristics of the new glass. Nearly 50 percent of the sunlight hitting the glass was reflected away from it. The resultant savings in air conditioning were considerable.

Though Bell Labs was the first building to utilize mirror glass, the first building totally constructed of the new glass was the headquarters for John Deere in Moline, Illinois (Fig. 5-11). It was completed in 1963. Saarinen said of it, "Having selected a site because of the beauty of nature, we were especially anxious to take advantage of views from the offices. To avoid curtains or venetian blinds, we worked out a system of sun shading with metal louvers and specified reflective glass to prevent glare."

The two buildings were dissimilar in appearance. Bell Labs was an elegant Miesien glass box. John Deere was more romantic, looking Japanese with its trabeated steel overhangs. Bell Labs was isolated from nature, an industrial Versailles in the midst of a formal landscape. John Deere was nestled in a tree-filled hollow. Even the glass was different. The cladding at Bell Labs, after all the dark tinted

Fig. 5-10. Bell Labs after expansion in 1967. Mirror glass was applied to the entire building. (*Courtesy Bell Labs*)

91

Fig. 5-11. Eero Saarinen and Associates: John Deere and Co., Moline, Illinois, 1963. A subtle mirror. (*Courtesy John Deere*)

glass was replaced, was silvery with a high degree of reflectivity. At John Deere, the reflectivity was cut back so that it was a slight haze on the glass. Also, the coating was gold in tint, as opposed to silver.

Bell and Deere were isolated experiments. Interestingly, industry as a whole had not been particularly intrigued with Saarinen's and Dinkeloo's tinkering. In the late 1950's, Dinkeloo had asked Libbey-Owens-Ford to help them with a full-scale mock-up of an office bay. According to Dinkeloo, the company refused. "We got the brush-off from the big boys," he said.

After John Deere was built, industry began to invest in plant and equipment to produce the new glass. The situation was reversed. Initially, architects had to prod industry to produce the glass. Now, industry had to sell the product to a much broader spectrum of architects.

The curtain wall, the building type that had made the interior mirror unnecessary, had created the conditions resulting in the development and acceptance of the biggest mirror of all, the mirror glass building.

Mirrors on the interior of buildings have a way of doing many things at once. They reflect people, space and light. Such is the case of the mirror on the outside of buildings.

Mirror glass was accepted rather rapidly by architects for a variety of reasons. By being reflective, it could reduce on air conditioning costs—but it could also make a building look "romantic" if it were near enough trees. By being opaque rather than transparent, it could make a building look more "modern." With the structure covered, the building would look like a sleek package rather than a boxlike cage. That package often came in varied, dynamic shapes.

At the time of the Manufacturer's Trust Building in 1954, buildings were supposed to be transparent. But just a decade later, architects were executing a 180-degree shift in their thinking. Suddenly, it was all right for buildings to be reflective.

Reflections of Images

In the early 1960's, glass companies started to sell other architects on the idea of reflective glass based on its ability to save on air conditioning costs. A building with a 44 percent reflective coating (similar to Bell Labs) could be expected to save two-thirds on air conditioning costs and a similar amount on the cost of the equipment itself. Installation of the equipment was also cheaper, with about a one-third savings there.

But in the 1960's, with abundant and cheap energy, architects were not particularly concerned about savings in that category. For instance, a building might be constructed with electric lights ganged together so that if one office were lighted, adjacent offices would be lighted as well, giving the building a strong identity at night.

Architects were initially attracted to the shiny new glass, not by messages of cost savings, but by discussions of aesthetics. Architects were sold on the new material because of the way it reflected images on its surface. A 1966 *Business Week* article said:

In trying to pry their way into the market, the glass companies are concentrating their sales talk on the mirror-like appearance of the glass. They are beaming their pitch at such architectural firms as Vincent G. Kling and Associates, which recently specified Kinney's deep gold glass for a new Monsanto building in Saint Louis. Says a Kling spokesman: 'We wanted the warmth of gold, and felt the glass was of our age. It gave the building a desired continuity with its surroundings.'

The article then went on to quote a spokesman from Kinney Glass who said,

We tried to sell architects on the economic advantage and paid no attention to the way the glass looked. It was as if the auto industry launched a new model sales campaign by completely de-emphasizing styling. Now we've discovered its aesthetic value, and we're selling.

The glass, which had the ability to make an entire building blend into its setting visually, came on the market at a time when the United States was entering an environmental movement. With space explorations and live broadcasts from the moon itself, people in the United States and indeed much of the world realized how small and insignificant the world was when viewed from outer space. With the awareness of the fragility of planet Earth came a desire to protect nature. Environmental legislation was passed to save endangered species and to set pollution controls.

The tired old curtain wall could look romantic and in tune with nature if it were clad in the new glass—and surrounded by enough trees. In a particularly amusing advertisement for reflective glass, a gingham-clad peasant girl is seated in a forest clearing with a basket of flowers. She is dreamily leaning against, of all things, a spectre of the corporate world, an office building. But, no problem. The office building has been reduced to the size of a tiny doll house. And the reflective glass skin dematerializes it all the more. The advertisement had two main messages in the text. The first was the reflective buildings would "reflect the natural grace and beauty of the environment around them." The second was that the glass would require less air conditioning, hence it would result in less pollution of the environment. In theory, such buildings would be less intrusive than buildings of the past, both visually and in terms of environmental pollution (Fig. 5-12).

As we will see of course, reality didn't always coincide with theory. These buildings were often glaringly noticeable, and heated up their immediate environments through reflecting sunlight.

THE OPAQUE SURFACE

These buildings were popular not only because of the re-

Fig. 5-12. Images of nature were important selling points for mirror glass. (*Courtesy Architectural Record*)

flections that they revealed, but also because of the various aspects of the building that were concealed by the seemingly opaque outer skin. Sometimes dynamic shapes emerged, sometimes the effect of streamlining. Sometimes a building could look extraordinarily lightweight, all because of this opacity.

Many had become tired of the 1950's style curtain wall building. One reason was that this building type was frequently boxlike in shape. Since the structure was revealed, the form of the building frequently derived from the regularity of that internal structure. But with the structure concealed by the shiny new skin, many shapes could result based more on architectural fancy rather than structural requirements. Hotels were often baroquely curving, offices were often sharply angular. Many of these office buildings in particular bore a strong resemblance to Mies van der Rohe's early sketches of the 1920's in which buildings were prismatic glass expressions, alternatively reflective and transparent, rather than structural expressions of his later work. Cesar Pelli has noted the "glassiness" of Mies' early work. He has also noted how greenhouse architecture was more an expression of glass shape rather than metallic structure.

Though these mirrored buildings had derived from an older tradition of glass expression, they looked new. The glass companies dubbed them "shaped glass." They did not seem to have been made in increments, but carved out of one large block of ice (Figs. 5-13 and 5-14).

For many of these buildings, since the surface was curving or angled, the resultant reflections on the surface were less important than a prismatic expression. They were shimmering objects in nature, more than smooth reflections of nature.

The shaped glass buildings were usually total glass expressions with minimal mullions that were almost invisible. Even their spandrels, the glass that covers the horizontal structure, were made of the same reflective material, although with a solid backing. The result of this uniformity of treatment was often a remote, totally abstract expression.

Fig. 5-14. Odell Associates Inc.: R.J. Reynolds Industries, Inc., Winston-Salem, North Carolina, 1976. Angular shapes were popular for offices. (*Courtesy PPG Industries, Inc.*)

Fig. 5-13. Welton Becket Associates: Reunion Center, Dallas, 1978. "Shaped glass." Hotels were often baroquely curving. Photo Balthazar Korab.

Fig. 5-15. Whirlpool Administration Center, Benton Harbor, Michigan. Transparency of 1/4" plate was often too revealing. (*Courtesy Libbey-Owens-Ford Co.*)

Sometimes designers combined the reflective glass with other materials. Perhaps the glass would constitute but one wall of a building. Very often, it was applied in horizontal ribbon windows, a continuation of the international style "skin," a taut, smooth membrane covering a volume.

In these instances, the other materials tended to give the building an identity and often a sense of scale. The combined forms were not as totally as abstract as the all-glass expressions.

In the ribbon window designs, the opacity of the glass was an important design element. For by being opaque, once again, the structure and interior clutter would be concealed. The only expression remaining was horizontal, sleek and streamlined.

A remodeling of the Whirlpool Administration Building is a case in point. In the original building, the transparency of the glass made the building look stodgy by revealing vertical columns and bunched, essentially vertical draperies. When the building was reglazed with mirror glass, the verticals were concealed by a totally horizontal expression. Now, the vertical lines did not slow the eye. The glass, along with a simplified mullion expression, made the building look more "modern," a smooth package which revealed none of the messiness of everyday life (Figs. 5-15 and 5-16).

Another trend in which the opacity of the mirror glass was exceedingly important was the cladding of entire traditional, usually masonry, buildings with mirror glass panels. Often, large parts of the cladding were spandrel glass, mirror glass with a solid backing. The result was to make an older, traditional building look like a new curtain wall building (Figs. 5-17 and 5-18).

In the past, architects had used mirrors on interiors as a means of dematerializing mass in architecture. Now, in these remodelings of older buildings, the same effect occurred. What had been a traditional bearing wall building, an expression of mass, now through the use of mirrors looked like a curtain wall expression of volume. It now looked lightweight instead of massive. One assumed that behind

Fig. 5-16. Whirlpool Administration Center after being reglazed with mirror glass. It is a simplified, totally horizontal expression. (*Courtesy Libbey-Owens-Ford Co.*)

Fig. 5-17. The Hancock Savings and Loan, Findlay, Ohio, just before renovation. (*Courtesy Libbey-Owens-Ford Co.*)

the skin of glass were slender columns supporting concrete slabs, rather than foot-thick walls of brick and mortar.

This exterior opacity has meant that spaces needing privacy can still have large glass windows, positioned somewhat more freely than would have been the case with clear glass. James Lambeth placed the master bedroom of his own house adjacent to the entrance. Privacy was assured since the room was glazed in mirror glass. This writer designed a poolhouse that required privacy in the dressing room areas. Yet, the pool itself was such a lovely view that the occupants of the dressing rooms could have both the view of the pool and privacy while they changed their clothes since mirror glass was installed at the windows. John Portman has juxtaposed hotel towers clad in mirror glass with office towers clad in regular glass. The hotel spaces still retain privacy, looking like mirrors from the outside. Of course, this seeming external mirrorlike opacity and internal transparency might result in voyeurism. Some office workers were questioned whether they liked a reflective solar film that had been installed on their ground floor office. They said they liked it. They remarked, "Why, a girl stood in front of that door for a half hour, combing her hair and putting on her make-up. We got a big laugh out of that."

Of course, when the lighting conditions are reversed, the opacity-transparency relationship is reversed. Suddenly, the interior becomes the mirror. Some have even considered this characteristic a desirable quality. Harry Weese wanted to install mirror glass on the windows of the subway cars in the Metro system he designed for Washington, D.C. When the cars were hurtling through darkened tunnels, people wouldn't look out and see unfinished concrete. Rather, they would see cheerful reflections of the bright cars. However, clear glass was installed in the cars instead.

Fig. 5-18. The Hancock Savings and Loan after being covered up by mirror glass. Instantly, a bearing wall building could look like a curtain wall building. (*Courtesy Libbey-Owens-Ford Co.*)

AMBIGUITY

Avant-garde architects experimenting with reflective surfaces in apartments and lofts in the late 1960's exploited
the ambiguous nature of the mirror. They used warped
mirrored surfaces so that one would not escape the dual
nature of the mirror, the fact that it was surface as well as
depth. This ambiguity was a part of the new mirror glass
buildings that were being built at the end of the 1960's.
They were just as ambiguous as the mirrored lofts, though
in a more subtle way.

C. Ray Smith, in his book *Supermannerism*, said that
"Mirror-faced buildings fascinated the minimalist architects since Saarinen's Bell Telephone Laboratories Building
in Holmdel, New Jersey (1962), but these did not aim for
ambiguity." Perhaps the aim was not, but the effect most
certainly was.

The mirror is itself an ambiguous material. And these
mirror glass buildings never appeared as other than mirrors.
They did not resemble pure space. They didn't fade away.
Even with minimal joints, one still saw the joints of the
glass. Reflections were often distorted and warped due to
imperfections in the heat-strengthened glass. The sun
bounced off those mirrored facades which was a give-away
of their reflective surface quality. Though not as icono-
clastic as those avant-garde interiors, they were just as
ambiguous.

These buildings were ambiguous also because of the
opaque quality of mirror glass. While architects had undoubtedly been drawn to the glass because of this quality,
it made for a lot of buildings that were difficult to interpret.
One couldn't tell what was behind the glass. Was it an
apartment building, a hotel or an office (Fig. 5-19)?

Of course, the international style had an element of

Fig. 5-19. Thompson, Ventulett and Stainback: Triangle Building, Atlanta, 1974. Unlike the magician's mirror, these buildings never disappear from our consciousnesses. (*Courtesy Libbey-Owens-Ford Co.*)

ambiguity inherent in its design. Since the building envelope was neutral, one often couldn't tell the difference between a chapel or a heating plant. This was the common criticism of Mies van der Rohe's Illinois Institute of Technology, for instance. However, if one merely looked through the transparent glass, one could gain more information and make a judgment. These buildings were never as ambiguous as the mirror glass buildings.

In the late 1960's, architectural magazines devoted a lot of coverage to what students and experimental architects were doing. The supergraphics, the bright colors, the silver surfaces, were seen, absorbed and often reinterpreted by the more establishment firms using more durable and expensive materials. Both groups were using reflective materials. Though their works looked rather different, one from the other, ambiguity was a common link.

OTHER MIRRORS

While architects were often utilizing highly reflective mirror glass, other architects exploited the reflective nature of tinted glass. Obviously, reflectivity was becoming acceptable, indoors, in mirror glass buildings, and in tinted glass buildings.

As noted before, clear glass contains the ability to create images on its surface, particularly if viewed obliquely and in dim light. Tinted glass, in concealing more of what is behind it, has even more of a mirrorlike quality.

Cesar Pelli was one such architect who worked to a large extent with tinted glass. He described his work in relation to theories of physical and perceptual transparency. Physical transparency would be in a sense "pure" transparency. One would see right through a material without ever realizing it was there, it would be so clear. Almost like the magician's mirror, the material would disappear on a perceptual basis. But the more common form of transparency would be perceptual transparency. One would perceive that one was indeed looking through a transparent

material, because one would be aware of the presence of that material. One would undoubtedly see reflections on that material. Clear glass, for instance, would permit views through it, but would also form other images on its surface under certain lighting conditions.

Pelli purposely chose tinted glass rather than clear glass so as to boost the perception of reflectivity in that glass. He described his approach in *Architecture and Urbanism*:

The type of glass used is also important. Tinted glass gives a better balance between transparency and reflectivity. Mirror glass (which I rarely use as it makes for a blind building) is almost totally nontransparent. Mirror glass can be modified so as to balance reflection with transparency, for example by striping as Kevin Roche used it in his elegant Irwin Union Bank in Columbus, Indiana. Clear glass on the other hand has very weak reflections. In few instances do the reflections overcome the light coming through.

Obviously, though Pelli rarely used mirror glass, the reflections that played across the surface of his architecture were an integral part of his work (Fig. 5-20).

In 1978, an article appeared in *The New Yorker* magazine written by Susan Lessard. Entitled "The Towers of Light," it conveyed her observations of the variety and changeableness of the reflections that played across the numerous glass skyscrapers in midtown Manhattan. However, when she interviewed architects, she found that they often ignored the reflections. She wrote, "Professional people I have spoken with, however, have found the very subject of reflections odd, possibly a little perverse." Some designers still thought of glass as a neutral material, more transparent than reflective. And when she interviewed I. M. Pei regarding the John Hancock Building, a mirror glass building, she was surprised at his response. He thought of the glass more as a taut smooth surface that concealed desks and venetian blinds, rather than an illusionary material constantly changing with varied lighting conditions. For

100

Pei's office, the planar aspects of the glass were more important, at least in the design stage, than the illusionary aspects.

As we have seen, and will see, by the end of the 1970's, there were many architects experimenting with tinted glasses and reflective glasses that were capable of producing illusions. Sometimes the illusions were at the forefront of their consideration. But very often, the illusions were a supplementary aspect of their designs. They realized that the building itself was finite and always visible, whereas the illusions would come and go, and change, sometimes looking glorious, sometimes gray and monotonous. Yet, the building would remain the major element on our consciousnesses.

We have all noticed how golden and shimmering the glass can look on a skyscraper at sunset. Yet, we differentiate between the building and the reflections. Even in a brilliant sunset, an ugly building will remain an ugly building.

ENERGY CRISIS

In all this discussion of illusionism and opacity and aesthetics, we forget the real *raison d'être* of mirror glass. It was developed initially as a means of reducing heat buildup in curtain wall buildings. Though Saarinen and others recognized the aesthetic potential in glass with a reflective coating, had there been no problem of heat buildup in all glass buildings, mirror glass would undoubtedly never have been developed.

After the Arab oil embargo of 1973, it suddenly became more expensive to heat and to cool buildings. For offices, typically, cooling was more expensive than heating. People, machines and lights, as well as incoming sunlight would produce heat in offices. This heat could be recycled in the

Fig. 5-20. Cesar Pelli for Gruen Associates: Pacific Center, Vancouver, British Columbia, 1968. A combination of reflectivity and transparency sought with tinted glass. (*Courtesy Cesar Pelli*)

101

Fig. 5-21. John Portman Associates: Hyatt Regency Hotel, Atlanta, 1971. Sometimes sun bounced off mirrored buildings in the form of nuisance reflections. Photo Jerry Spearman.

winter, but in the summer it had to be replaced with cool air. For this reason, there was an increased emphasis on blocking heat-producing sunlight before it ever reached the building interior. Tinted and reflective glasses became all the more prevalent.

A 1974 *Progressive Architecture* article said: "In response to the energy crisis, they [architects] have strained to capacity the manufacturers' ability to make various types of insulated and reflective glass." Just 12 years after the construction of Bell Labs, industry once again lagged behind architects' requirements for new glasses. By 1973, the order of magnitude was much greater. By this time, there was considerable amount of variation in the types of reflective glass on the market. The color of the reflective coating could vary, as could the color of the glass. The coating could be bright, or almost nonexistent. The heaviest coating would be about 44 percent reflectance, with only a 7 percent daylight transmittance. Other lighter coatings would allow approximately a 30 percent daylight transmittance. Since reflective glass spandrels often looked identical to the vision glass, the amount of vision glass could be reduced while retaining a total glass expression. In actuality, these buildings were typically 30 percent vision glass.

Solar films also came on the market in the 1970's. They were plastic films with a reflective coating which were applied to the inside of existing windows. These solar films were subject to scratching from the back by cleaning products, and occasionally would buckle. Yet, in spite of various problems, their use continued to grow, particularly on west- and south-facing facades.

So many mirrorlike walls began to appear in the United States that by 1976, New York Times architecture critic Ada Louise Huxtable termed the mirror glass building "indisputably the building of the year, ranging from elegant to awful and spreading like a funny house plague in cities and limbo. It reflects streets, freeways, buildings or no buildings, heat and glare, and is the latest architectural copout."

PROBLEMS WITH THE MIRROR GLASS BUILDING

Mirror glass was developed and accepted because of technical and aesthetic problems with the glass curtain wall building. However, mirror glass buildings contained plenty of their own problems as well, technical and aesthetic.

As for heat buildup, mirror glass eliminated it on the insides of buildings by reflecting away sunlight. However, that extra energy didn't just disappear. It heated up the environment just outside the mirror glass building. In some instances, the heating loads of neighboring buildings were doubled. This resulted in the owners of older buildings taking the owners of mirror glass buildings to court and suing for damages.

Interestingly, when reflective glass was first being marketed, some industry spokesmen did not recommend the material for urban use. A 1966 *Business Week* article said,

Reflecting glass is not the fabric of which cities are made, however. Kinney recommends that its Reflectovue glass be used on open landscapes where it will reflect trees and sky. One reason for this, Bleecker admits, is that if the glass sheathed skyscraper-lined streets, it would reflect so much heat pedestrians would find the temperature oppressive.

Nuisance reflections of direct beams of sunlight were also a problem with the shiny glass. The more reflective the glass, the more light could bounce off the walls of the building and into the eyes of unsuspecting motorists. The problem was usually most bothersome when the sun was relatively low in the sky, around rush hour (Fig. 5-21).

Reflections of images were often less attractive than had been depicted in advertisements (Fig. 5-22). Cars, television antennas and other unattractive buildings were all too often the items duplicated. In Washington, D.C., a mirror glass building was situated directly across the street from a yellow brick parking garage with square, unadorned,

Fig. 5-22. Images on mirrored facades were not always of trees and clouds. (*Courtesy Libbey-Owens-Ford Co.*)

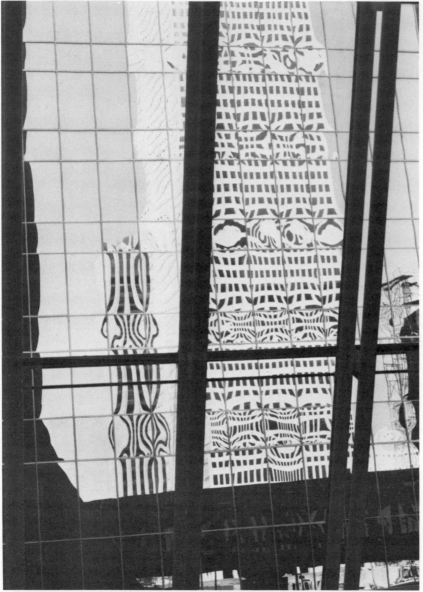

Fig. 5-23. Distorted images were a common aesthetic problem, particularly in the spandrel areas. Photo Pamela Heyne.

unaligned windows. That parking garage was the sort of structure one would like to ignore, yet it was duplicated in the facade of the mirror glass building. The mirror, more than any other material, depends on its immediate environment for its appearance.

Urban reflections, besides being unattractive, were also illogical at times. Rapidly moving cars reflected on a facade gave a building a curiously kinetic appearance. The reflections of other buildings sometimes were strange. The reflected facades, of structures say across the street, were naturally smaller than facades adjacent to the mirrored building. This sometimes resulted in a scale problem in the streetscape. Also, the reflections were often quite distorted. While this was not objectionable when sky and trees were reflected, it was odd to see images of crumpled-looking buildings.

Ordinary float glass, when mirrored, had minimal distortions. But this glass was usually not considered safe enough for exterior use, particularly in high-rise construction. Heat-strengthened or tempered glasses were often specified. These glasses had been heated again after leaving the annealing oven, resulting in some surface warping. Tempered glass was the stronger of the two types, and the most distorted.

While most high rises were constructed of heat-strengthened glass in the vision areas, very often tempered glass was placed in the spandrel areas because of the problem of heat entrapment there, often resulting in great temperature differentials between indoors and out. Some tried to minimize the aesthetic problem by specifying that all exterior spandrel panels be convex glass which has less of a "funhouse" quality than concave. But an aesthetic problem still existed (Fig. 5-23).

Then there was the difficulty, particularly in taller buildings, of changing conditions of clouds and sky. The same building that would convey images of bright blue sky and flowing clouds in June, would look gray and sodden on a cold November afternoon (Figs. 5-24 and 5-25).

Fig. 5-24. Welton Becket Associates: Valley Center, Phoenix, 1973. Reflections of the sky could be delightful. Photo Marvin Rand.

Fig. 5-25. Valley Center. Reflections could also be oppressive. Photo Marvin Rand.

105

Fig. 5-26. K.R. Cooper: offices for Ontario Hydro, Toronto, 1976. Scale, particularly in urban areas, was difficult to resolve for mirrored buildings. (*Courtesy Libbey-Ownes-Ford Co.*)

One of the most serious aesthetic problems with the mirror glass building was that of scale. With the development of minimal mullion systems, these new structures often looked as blank and scaleless as a glacier (Fig. 5-26). The mullions were often disposed in a regular fashion so that one could not tell where the floor was and where the window was. Arthur Drexler, writing in the Museum of Modern Art's catalogue for the show, "Transformations in Modern Architecture" said, "The more regular the grid, the more abstract the elevations. Perfect regularity results in perfect meaninglessness, except that the whole conveys an indifference to human presence most people interpret as hostile."

Another problem of these buildings occurred on their interiors. Since light was reflected away from their facades, that light would not be transmitted through the glass. Hence, from a psychological point of view, the view to the out-of-doors could be dark and depressing. Kevin Roche has noted this problem, particularly with glasses of high reflectance, and conversely, low transmittance.

In 1975, the Department of Commerce compiled a survey of available literature on the psychological implications of windows. The impetus behind this study was the energy crisis. Though some critics felt windows were excessively wasteful of energy, others felt that they shouldn't be a scapegoat because they were necessary from a psychological point of view. The point of the study was to assess this psychological importance. Entitled *Windows and People: A Literature Survey*, it came to the conclusion, not surprisingly, that people liked windows. They imparted a dynamic quality to the interior environment and gave that environment a sense of spaciousness. Also, the window per se was more important than the particular view from that window. When people were deprived of windows, their psychological well-being was adversely affected. Factory workers, when deprived of views by say, tinted blue glass, were known to break the glass in order to gain more contact with the out-of-doors.

The only interiors in which people did not need windows, and did not feel deprived by not having access to them, were facilities such as theaters and department stores. Evidently, there was already enough dynamism in such environments.

Some of the studies dealt with how people reacted to tinted and reflective glass from the inside looking out. With a high degree of light transmittance (over 50 percent), people had little negative reaction. But, in highly reflective glass with a small degree of light transmittance (12 to 15 percent), some negative responses were cited:

The occupants of the rooms complained of the depressing visual effect of the external view due to the reduction of luminance, the need to use artificial light during daylight hours in a room with the external wall 60 percent glazed and the fact that the glass became a mirror to the room interior when external illumination levels were low. The effect, which was intermittent in daytime, resulted in peripheral visual distractions to the occupant.

From an energy standpoint, this reduction in luminance could also prove a problem. It prevented a certain amount of useful solar gain, possibly desirable in winter months. Also, with less natural light, there was more need to use electric lights in the space.

The ability to keep buildings cooler, to reflect images, to create an external opacity and to avoid internal shading devices were all reasons architects started using reflective glasses and films. But when improperly used, these same factors became enormous problems for the immediate environment and for building occupants.

THE IMPORTANCE OF SAARINEN

Despite the litany of problems with mirror glass, many architects have handled it with understanding. They have

*p. 66, *Windows and People: A Literature Survey*.

avoided potential surface blankness with careful manipulation of that surface. They have also realized the material can be bright, reflecting images and sunlight, or subtle and softly shimmering.

Though Saarinen only built two mirror glass buildings before his death, he understood the potential and limitations of the material. Bell Labs, for instance, was a bright mirror, yet the reflections on its surface were aesthetic. Set in the middle of a vast meadow, it reflected greensward, man-made lake, Saarinen-designed water tower, and distant trees and hills. The vast parking lots for the facility were located on the short sides of the building, and were screened from view by coniferous trees. Loading docks, potentially unsightly items, were located underground. At John Deere, the glass didn't reflect images, but was a soft, subtle haze.

Interestingly, a number of the architects who have worked successfully with mirror glass apprenticed in their early years with Saarinen. Kevin Roche and John Dinkeloo of course continued his office after his death. They experimented with alternating clear and varying degrees of reflective coatings on glass. They took mirror glass out of the vertical plane and used it as a canopy, hovering protectively over a clear glass window. They combined mirror glass with different materials.

Other architects who worked with Saarinen included Cesar Pelli, Tony Lumsden, Paul Kennon and Gunnar Birkerts. After leaving Saarinen's office, they experimented with mirrors and mirror glass in diverse ways. Cesar Pelli's California Jewelry Mart is as bright and playful as Christmas tree ornaments; his American Embassy in Tokyo is serene. Tony Lumsden's design for the Lugano Convention Center in Lugano, Switzerland, seemed the ultimate in curving shaped glass expression. Paul Kennon juxtaposed the Bell Telephone Laboratories Building in Columbus, Indiana with a monumentally scaled trellis to compensate for the reflections in the immediate urban environment. Gunnar Birkerts was the first designer to install a periscope in an architectural fashion on a building.

Fig. 5-27. Eero Saarinen and Associates: Bell Labs, Holmdel, New Jersey, 1962–1967. Reflections on the facade were carefully considered. (*Courtesy Bell Labs*)

Saarinen was interested in glass, thinking it the material of our age. He also had a vast curiosity, a need to solve new problems in new ways, and a need to understand projects thoroughly before they were built. His designs and his approach obviously attracted some of the best architects to work with him initially, and provided them with a good foundation for their later work.

Many were in the office when mirror glass was first being developed. And many used mirrors in an everyday way in conjunction with the numerous models that Saarinen demanded. Sometimes just half a model would be built, with the rest completed by a mirror. And sometimes they would "get into" the models with miniature periscopes. These handmade periscopes helped them understand the model at the scale of the model, and seemingly walk around inside.

When Kevin Roche was asked if at the time of Bell Labs it was at all frightening to be using mirror glass, a material that architects had never used on a building at such a large scale, he replied, "It didn't seem so at the time." Obviously, Saarinen and the architects working with him realized that while the mirrors at Bell Labs were keeping the interior cooler, they would be reflecting well-considered images. Bell Labs did not suffer from the haphazard quality all too common in mirror glass buildings (Fig. 5-27).

SOLAR MIRRORS

The energy crisis spurred the development of the mirror glass building. It also spurred the growth of the solar mirror. So, a little more than a decade after Bell Labs, mirrors were being investigated and used as one means of obtaining energy from the sun. They were one aspect of solar energy.

Prior to the oil embargo, most interest in solar energy, including mirrors, came from backyard tinkerers, isolated scientists and a few visionary architects. But with the cutoff of a major source of imported oil, Americans realized that petroleum products were rapidly being used up. Their own petroleum supplies were increasingly expensive to retrieve. Imported oil was subject to sudden price escalations. With varying political situations, as the embargo demonstrated, supplies were not always reliable.

Suddenly, the nation began to investigate alternate energy sources, such as coal, nuclear energy, synthetic fuels and solar energy. The government invested money in numerous areas simultaneously. Much of the investigation was spearheaded by the newly formed Department of Energy.

The Department of Energy's definition of solar energy was a broad one, covering eight categories divided into three major groups:

1. Thermal Applications
   Heating and cooling of buildings, including hot water heating
   Agricultural and industrial process heating
2. Fuels from Biomass
   Plant matter, including wood and waste, sea algae
3. Solar Electric
   Solar thermal electric, such as the power tower
   Photovoltaics—solar cells
   Wind—windmills
   Ocean thermal electric
   Hydropower—hydroelectric dams

Solar mirrors straddled two of the three groups. They were used for thermal applications and the creation of electricity. By focusing sunlight, they could create varying degrees of heat, which had many practical applications.

Solar mirrors do not somehow mysteriously create new energy. All they do is redirect sunlight. They focus it along a line, or to a small point. This is important, because although sunlight is infinitely renewable, plentiful, and available to all nations, it is diffuse. To create temperatures of more than, say, 180 degrees Fahrenheit, it needs to be concentrated. This has been the objective of solar mirrors for centuries.

Mirrors reflect sunlight. When light strikes the appropriate medium, it is transformed into usable heat. For this reason, solar mirrors even work in a northern setting like Minneapolis. As long as enough sun is shining, its energy can be tapped through concentrating it. Depending on quantity and design of mirrors, a wide range of temperatures can be produced.

Perhaps the most dramatic example of solar mirrors in the world today is the giant solar furnace at Odeillo, high in the French Pyrenees. Odeillo consists of tracking mirrors that redirect sunlight to a parabolic mirrored wall, a curved shield of 12,000 mirrors. The parabolic wall concentrates that light toward a 2' in diameter aperture in a tower placed at its focus. The temperatures in the tower can reach 7000 degrees Fahrenheit. The purpose of such extreme heat is to melt alloys without contamination. Some of that energy has been tapped to generate electricity for the surrounding region (Figs. 5-28 and 5-29, Plate 27).

In the United States, high-temperature designs have been used to generate electricity. The "power tower" is typically several stories high, surrounded by mirrors that track the sun. The mirrors would focus sunlight toward a boiler at the top of the tower, creating temperatures in the 2000 degrees Fahrenheit range (Fig. 5-30). This would produce steam to run turbines and electric generators. Another experimental form might be a 200' in diameter mirrored dish. It would remain stationary while the boiler on a boom would track the sun. Temperatures here might be 1000 degrees Fahrenheit, and electricity would also be a product. Other forms involve photovoltaics, or power cells. These are thin slivers of silicon, treated to interact with sunlight and create electricity directly, without requiring the steam process. These are often more efficient with concentrated sunlight. Though Fresnel lenses, similar to convex pieces of glass, are frequently combined with the

Fig. 5-28. A solar furnace near Odeillo, France, designed to create temperatures up to 7000 degrees Fahrenheit through focusing mirrors. (*Courtesy Department of Energy*)

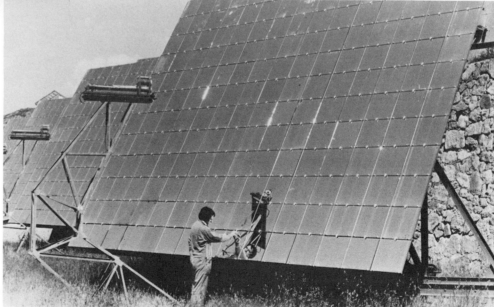

Fig. 5-29. A heliostat, or tracking mirror, at Odeillo. (*Courtesy Department of Energy*)

solar cells, in some instances mirrors have been used. To date, the solar cell, though intriguing for its simplicity, and for the fact that silicon is among the most plentiful substances on earth, is still experimental and highly expensive (Fig. 5-31).

Medium-temperature reflectors are used to heat and cool buildings, to pump water for agricultural purposes and to create steam for various industrial processes. One of America's favorite foods, the french-fried potato, has been prepared in a plant using oil heated by solar energy. The designs are varied. There are parabolic concentrators, compound parabolic concentrators and dish-shaped concentrators, to name a few. Their operating temperatures range from around 200 to 400 degrees Fahrenheit and beyond. Some designs track the sun, some are seasonally tilted to allow for varying solar angles, and others are stationary. Since the early 1970's, a number of companies have emerged that produce these reflectors. Some can be obtained by architects merely by flipping through Sweet's Catalogue, the architects' source book (Fig. 5-32).

Then there are various low-temperature reflectors. These might be simply a canted mirror glass window, designed to redirect sunlight to a swimming pool below, or perhaps a movable reflective flap adjacent to a southern window. They reflect sunlight upwards, downwards and sideways, wherever a little extra energy is needed. One of the simplest versions of this type of reflector is the so-called "bread box." A favorite of backyard builders, it has been around since the 1930's, and is basically a glass-enclosed box containing a water storage tank. The backing and movable top of the box are covered in reflective materials to intensify heat contained in the boiler. In 1975, an 84-gallon system was installed at the Farallones Institute Rural Center in Occidental, California. Throughout that winter, it produced water temperatures above 130 degrees Fahrenheit. That was enough energy to provide hot showers for 15 students and staff. While not attractive, it demonstrates that reflectors are not necessarily complex (Fig. 5-33).

Fig. 5-30. A power tower, looking like the backdrop of a science fiction movie, creates electricity from steam which has been heated by heliostats. (*Courtesy Boeing*)

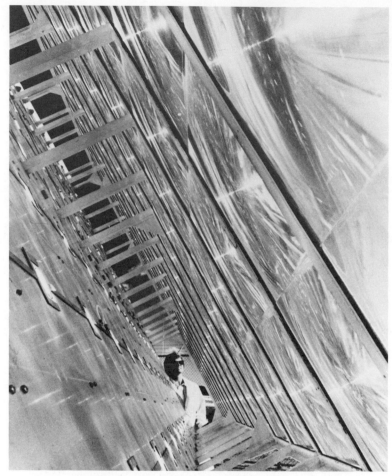

Fig. 5-31. Photovoltaic cells, used here in conjunction with Fresnel lenses that intensify sunlight. Occasionally, mirrors have been used to increase the amount of sunlight striking these cells. (*Courtesy Sandia Laboratory*)

Fig. 5-32. A bank of parabolic reflectors to operate at medium-temperature levels. In this picture, the receiver rod has not been installed. (*Courtesy Sandia Laboratory*)

In spite of the wide variety of reflectors, some mass-produced, some custom-made, they are not the most common form of obtaining energy from the sun today. The most prevalent approaches to solar are passive designs and the flat plate collector.

Typically, in passive design, the building acts as a giant solar collector. It absorbs heat during the day, and gradually releases it during the nighttime hours. Heat is transferred from room to room by radiation, conduction and convection. In essence, passive design makes a virtue out of the greenhouse effect. Sunlight from the exterior changes form on the interior, warming that interior. (To further complicate things, some passive designs incorporate reflectors or flat plate collectors.)

Typically, the flat plate collector is a glass-enclosed box mounted on a south-facing roof. In it are tubes which contain some medium to be heated, such as water or air. These collectors absorb energy as it comes directly from the sun. Sunlight is not concentrated. Their operating temperatures range from 100 to the maximum of 200 degrees Fahrenheit. This is adequate for domestic hot water (bathing and laundry) and space heating. However, if one desires a cooling system to operate in conjunction with the other functions, the temperatures are too low. While there are various passive auxiliary cooling systems available, absorption air conditioning is to date the most efficient means of cooling with solar energy, particularly large structures. Curiously, it needs a lot of heat in order to cool.

Absorption cooling devices work best with fluid temperatures between 250 and 300 degrees Fahrenheit. The lowest possible fluid temperature that can be used is 180 degrees Fahrenheit. At this level, the cooling devices have sharply reduced efficiencies.

Absorption air conditioners are similar to window air conditioners and heat pumps. It's just that they don't need vast amounts of electricity for their operation. All they need is heat. They use two working fluids, an absorbant, such as water, and a refrigerant, such as ammonia.

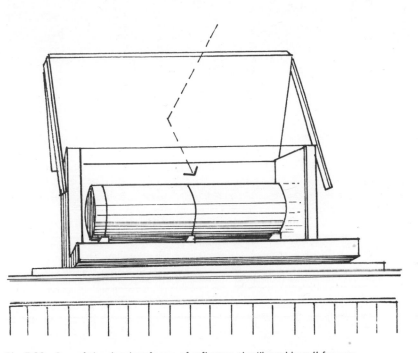

Fig. 5-33. One of the simplest forms of reflector, the "bread box," focuses energy toward a roof-mounted water tank. Not beautiful, it has been used by backyard builders since the 1930's.

Initially, the ammonia and the water are in solution. Then, heat from the solar collection system warms the solution so that the ammonia is evaporated. The ammonia then condenses, passing through the room via cooling coils. It evaporates again, absorbing the heat from the room. Eventually, the ammonia returns to the original water solution. Since most clients today couldn't live without air conditioning, it is likely that many architects involved with solar energy will utilize some form of reflector in their careers.

At first glance, solar mirrors seem complicated. But they are merely a dynamic way of using the mirror. As we have seen on the interior, the magician used the mirror in

Fig. 5-34. James Lambeth: Yocum Lodge, Snowmass at Aspen, Colorado, 1972. This reflector is integrated into the facade of the building.

a dynamic way so that the reflected ray of light from an observer would not return to his eyes. Thus, he created the effect of transparency. Other designers have shifted the reflected ray to create various periscoping devices. Now architects and engineers are shifting rays of light from the sun, and focusing those rays, so that a number of useful tasks can be performed.

Of course, solar mirrors are not panaceas. They are still expensive and experimental. And, like flat plate collectors, they require other forms of energy as a backup for periods when the sun is not shining. Some worry about the problem of stray reflections, theorizing for instance that the tracking mirrors around power towers could have a blinding effect on airline pilots. At the solar furnace installation at Odeillo, floor boards in nearby houses were burned due to misaligned mirrors. There are even some aesthetic problems with solar mirrors. Particularly in large-scale installations, the metal structure under the mirrors can resemble a busy forest of struts. When installed on a building, they sometimes look "tacked on."

In spite of the problems with the solar mirror, it is fascinating. It can perform a variety of tasks, is nonpolluting, and is very often beautiful. While it is reflecting energy, it can reflect dazzling views of the blue sky to our eyes. While it reaches for the sun, we often see new architectural and engineering forms emerge (Fig. 5-34).

For architects, the solar mirror is by far the newest way of using the mirror. It is not as ubiquitous as the mirror glass building, for instance. Indeed, it is still more of an engineering tool than an architectural one. In a way, this situation reminds one of other periods when architects started using a new technology. For instance, at the turn of the century, concrete was primarily an engineering material. A few architects, such as Auguste Perret, began to experiment with it. Now of course, we take it for granted as a standard part of the architectural vocabulary.

We are at something of a crossroads with the solar mirror. It may remain experimental, or it may become something that architects design and specify with ease.

6
Glass Expressions

Although glass technology may develop further—to explore color, texture and transparency
as well as reflections—at its present stage, architects have preferred to concentrate on a building's shape.

Arthur Drexler,
Transformations in Modern Architecture, 1979

A department store column that is covered in mirror doesn't really disappear. The mirror makes the column less dominant visually. But it always remains a mirrored surface.

When the mirror is out-of-doors, covering an entire building, it also does not disappear. The mirror is clearly covering a large object of some sort; one isn't always sure just what. That object could be a volume, a space defined by the glass alone. Or it could be a massive structure, like a bearing wall building.

It is fairly easy to ignore the mirror on the department store column. It is less easy to ignore the mirrored building, for two major reasons: brightness and scale. Since the glass out-of-doors is reflecting the bright sky as well as the sun itself, it is often a blinding apparition, continually flashing into our consciousness. Also, since the first mirrored buildings, shapes have become more angular, faceted, or comprised of multiple curves. These new forms, "shaped glass," sometimes reflect light the way disco mirrors do. As for scale, many mirrored buildings have a uniform exterior mullion treatment, making them particularly difficult to fathom. One finds oneself constantly staring at some of these buildings, trying to figure out just what is beyond those bright remote walls. Big Brother is watching you, but from exactly which window?

SUBTLE MIRRORS

The material itself is available in varying degrees of reflectivity. The reflective coatings can be applied to clear or tinted glass, on interior surfaces or exterior surfaces. The coatings themselves can vary in terms of color. With glasses of lower reflectance (and conversely, greater light transmittance), images on the surface will be hazier, and there will be gentle views into the building.

At times, many designers prefer the less reflective end of the mirror glass spectrum, using it as a softly shimmering surface, or combining it with clear glass. For some architects, the reflection of the surrounding environment per se is a relatively insignificant aspect of their design.

Philip Johnson, who co-authored *The International Style* in 1932, some 40 years later, discussed his design approach to mirror glass. He was quick to realize that it had to be handled differently from clear or tinted glass. Writing about the IDS Center in the November, 1973 issue of *Architectural Forum,* he said:

Mirror glass reflects so much more than ordinary glass as to constitute a basically new material. Some may not care for the effect, but there is no escaping the future of this new glass. And it is so much cheaper!

How, therefore, to handle it?

We felt this glass would make today's much despised 'glass box' even more boring than the usual speculative skyscraper; by not seeing through the glass but only at it would we get a monolithic, windowless effect which would be scaleless. Therefore, we used as transparent a glass as we could afford (20 percent daylight transmittance— the material is available with transmittance as low as eight percent). Second, we designed deeply projecting mullions and muntins combined with small panes of glass (30 inch on center verticals instead of the more usual 60 inch) to create a network of lines—more the aspect of a birdcage than of a glass box.

"But the most important are the zogs. The 90-degree zig-zag of the wall surfaces results in a range of self reflections which make dark vertical bands on the tower, relieving the great mirror surfaces. The zogs also have the delightful dividend of making a crisscross of spandrel bands at dizzying angles to one another.

The crisscross pattern on the glass was in essence a large-scale version of *glaces à répétitions.*

The IDS Center is an urbane complex which does not overwhelm with its presence. At street level, it is a welcoming building with clear glass entrances and views of its central lighted court. Shops are also integrated into the streetscape (Figs. 6-1 and 6-2).

Though Johnson-Burgee's design for PPG Headquarters looks rather bizarre, being the first high-rise, neo-Gothic

Fig. 6-1. Philip Johnson and John Burgee, Edward F. Baker Associates: IDS Center, Minneapolis, 1972. One does not see a smooth reflection of clouds and sky, but an articulated surface of low reflectivity. Photo Nathaniel Lieberman, Todd A. Watts.

mirror glass complex in the history of architecture, in many respects it is similar to the IDS Center. Its historicizing elements occur at the top and bottom of the building. In between, its treatment is similar to IDS, with glass of low reflectivity, closely spaced mullions and modulated surface. At the street level, it provides access to and views of open courtyards (Fig. 6-3).

Arthur Drexler, in the catalogue for the exhibit "Transformations in Modern Architecture," indicated that many of the unusual shapes of mirror buildings were made all the more startling in that they apparently derived from masonry forms. This is certainly true in the Gothic peaks and arches on PPG Headquarters. However, many forms that resemble pyramids or battered walls derive perhaps less from masonry prototypes than from the current fondness for large, greenhouselike spaces. Many of these sloping walls are treated with reflective materials to reduce solar gain. In Johnson-Burgee's Garden Grove Community Church, the architects sought an open yet mysterious environment. According to Johnson, "The Reverend Robert Harold Schuller believes God should be enjoyed in the presence of the sky and the surrounding world, not in the forbidding stone environment. The glass itself will be reflective glass which allows only 14 percent of the heat and light to penetrate. The atmosphere will be subaqueous" (Fig. 6-4).

Pennzoil Place in Houston is an example of shaped glass. It consists of two trapezoidal, pitched roof towers linked by glass roofed courts. The glass on the towers is of such a low degree of reflectivity and such a dark tint, that the building does not seem to be a mirror glass building. The most reflective glass is used on the angled mirror glass courts. It is a building that is dynamic and subtle at the same time (Figs. 6-5, 6-6 and 6-7).

Another subtle mirror glass building is the John Hancock Tower in Boston, designed by I. M. Pei. Before this writer had thought much about the nature of the mirror out-of-doors, she questioned the partner in charge of design,

Fig. 6-2. IDS Center. The complex is transparent and welcoming at the street level. *Glaces à répétitions* was a technique used by Johnson and Burgee to add interest to the glass.

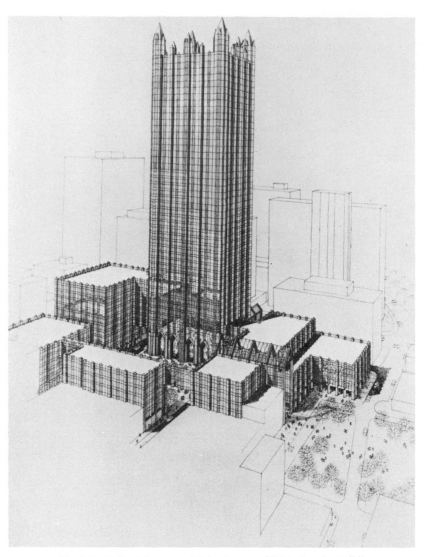

Fig. 6-4. Philip Johnson and John Burgee: Garden Grove Community Church, Garden Grove, California, 1980. A cathedral of muted glass. Photo of model Louis Checkman.

Fig. 6-3. Philip Johnson and John Burgee: PPG Industries Building, Pittsburgh project. In spite of curious Gothic detailing, the treatment of the glass derives from IDS. (*Courtesy PPG Industries*)

118

Fig. 6-5. Philip Johnson and John Burgee: Pennzoil Place, Houston, 1977. The two trapezoidal towers are barely reflective. They appear merely to be tinted glass. Photo Forster Studio.

Fig. 6-6. Philip Johnson and John Burgee: Pennzoil Place, Houston, 1977. A brighter reflective glass at entrance. Photo Richard W. Payne.

119

Henry Cobb, about why mirror glass was used on the building. The question: "Did you clad the building in mirror glass so that it would disappear?" Cobb, of course, realizing that mirror glass would never really disappear out-of-doors in the glare of the sun, merely shook his head, mused a bit and said it was much more complicated than that. Cobb said, "The glass would be nothing without the form—it would be another stupid mirror glass building." In what was something of a surprise, he said that the building could almost have been made of concrete.

John Hancock was an enormous, New York-scale sky-scraper set in the midst of a fragile urban environment, Boston's Copley Square, bounded by H. H. Richardson's Trinity Church, Copley Plaza Hotel and the Boston Public Library. The architect's task was to design Hancock in such a manner that its enormous bulk would not overwhelm the nineteenth-century structures in the square (Figs. 6-8 and 6-9, Plate 13).

According to Cobb, the most important part in minimizing the bulk of Hancock was to skew the building slightly. Instead of being rectangular, the building is actually a rhomboid, with the upper stories angling away from the square. Hence, when one is in the square, one is primarily aware of the narrow face of the tower in the corner of the space, a traditional location for towers (Figs. 6-10 and 6-11).

The architects decided to bring the face of the building flush down to the ground, feeling that a colonnade would have competed with Trinity Church (Fig. 6-12).

Mirror glass was selected as the cladding of the building, not for its illusionary qualities, and not because it would mirror and duplicate, and hence defer to the surrounding buildings. It was chosen because it would retain the surface of building. I. M. Pei, quoted in the July 10, 1978 edition of *The New Yorker*, said regarding Hancock, "Anyone who chooses mirror glass for its reflections misses the point. The point of mirror glass is that it preserves the surface. Clear

Fig. 6-7. Philip Johnson and John Burgee: Pennzoil Place, Houston, 1977. Glass at entrance reduces solar gain from above.

Fig. 6-8. I. M. Pei and Partners, Architects; Henry N. Cobb, design partner: John Hancock Tower, Boston, 1976. A brightly shining mirror would have overwhelmed the setting. Photo George Cserna.

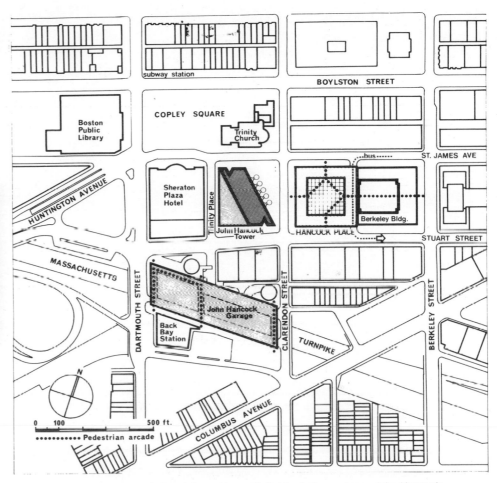

Fig. 6-9. I.M. Pei and Partners, Archtiects; Henry N. Cobb, design partner: John Hancock Tower and map of surrounding site. The narrow end of the tower faces Trinity Church and Copley Square.

121

Fig. 6-11. I. M. Pei and Partners, Architects; Henry N. Cobb, design partner: John Hancock Tower, Boston, 1976. Note how upper part of tower angles away from the square so that its mass will be diminished. Photo George Cserna.

Fig. 6-10. I. M. Pei and Partners, Architects; Henry N. Cobb, design partner: John Hancock Tower, Boston, 1976. Photo Gorchev and Gorchev.

122

Fig. 6-12. I. M. Pei and Partners, Architects: Henry N. Cobb, design partner: John Hancock Tower, Boston, 1976. Linear banding on Trinity Church, and on facade of the glass tower. Photo Gorchev and Gorchev.

Fig. 6-13. Kevin Roche, John Dinkeloo and Associates: Worcester County National Bank, Worcester, Mass., 1974. Subtle and bright reflective glass are juxtaposed.

123

Fig. 6-15. Kevin Roche, John Dinkeloo and Associates: United Nations Plaza Hotel and Office Building, New York, 1975. A mirror glass canopy hovering over clear glass windows.

Fig. 6-14. Kevin Roche, John Dinkeloo and Associates: United Nations Plaza Hotel and Office Building, New York, 1975. The building is a barely reflective bottle-green glass, chosen to blend in with the U.N. Secretariat.

glass is transparent. You see what's inside—unsightly venetian blinds, office furniture, and so forth."

The narrow facade on Copley Square looks all the more fragile, all the more planar, since one does not see definitely how large a building looms behind the facade. Basically, the glass stops your vision from going any further. It is an expression of skin. Yet, it is not completely abstract, for when close to it, one has a sense of human activity beyond, for the glass selected had a subtle reflectivity and tint. The building is also given scale, much like the IDS Center. The floor and column lines are defined by anodized aluminum frames, interspersed with lightweight, closely spaced mullions.

Since Kevin Roche finds looking through highly reflective glass depressing, one frequently sees him using clear or barely reflective glass in vision areas, more reflective glass at head and sill areas. Worcester County National Bank Building in Worcester, Massachusetts, featured a layered effect, with the least reflective glass being vision glass (Fig. 6-13). Roche/Dinkeloo's U.N. Plaza also features glass of varying degrees of reflectivity, with everything given a slight, bottle-green tint to match the U.N. Secretariat across the street (Figs. 6-14 and 6-15). Roche felt the U.N. Plaza was superior to Worcester because its shimmering glass had been "tied down" visually with a definite mullion system. He said, "There is a fine scale grid over the whole reflective glass building. As you move away, the grid reads as a surface. Philip Johnson did the same thing at IDS."

The Irwin Union Bank and Trust Company in Columbus, Indiana features narrow stripes of reflective coating applied to clear glass. The light on the interior has a variegated quality, as if filtered through a lattice (Fig. 6-16).

Richardson-Merrell is a particularly intriguing building. Its interior has mirror glass tables. Its exterior has mirror glass skylights, and a mirror glass canopy. The canopy shades the window glass, which is absolutely clear, from a buildup of heat and glare. Though mirror glass was first

Fig. 6-16. Keven Roche, John Dinkeloo and Associates: Irwin Union Bank and Trust Co., Columbus, Indiana, 1972. The glass alternates between clear and reflective stripes.

125

Fig. 6-17. Kevin Roche, John Dinkeloo and Associates: Richardson-Merrell Inc. Headquarters, Wilton, Conn., 1974. Clear glass through which to see a magnificent natural setting; heat gain has been controlled with mirror glass canopies.

developed by Saarinen as a means of combatting heat and glare, it was initially used in the plane of the wall. Roche/Dinkeloo accomplished the same thing by using mirror glass just beyond the plane of the wall. This permitted unobstructed views of the surrounding Connecticut forest. Heat was controlled, but not at the expense of the view (Figs. 6-17 and 6-18).

An intriguing use of mirror glass was shown by Cesar Pelli in a theoretical project. It was an "ideal house" to be exhibited in the Venice Biennale in 1976. It consisted of a long stone corridor and a series of glass spur walls, perpendicular to that corridor. On one side, the glass walls ranged from clear through soft rainbow tints. On the other side of the corridor, the glass partitions ranged from clear to softly mirrored effects. The glass partitions penetrated nature. They represented the poetry of architecture; one could look through the partitions and gain a misty, slightly evanescent view of nature. Pelli demonstrated with further drawings that real societal needs would ultimately intrude on the setting. Solid cubicles, representing rooms, would be inserted between the glass partitions, ultimately destroying the misty views. Though the project was theoretical, Pelli did consider ways of constructing it. Ultimately, he thought that the glass could be hung from beams (Fig. 6-19).

BRIGHTER MIRRORS

Cesar Pelli's California Jewelry Mart remodeling used mirrors as a means of brightening up what had formerly been a dark lightwell. Supergraphics were painted on the walls, and mirrors and painted plywood panels were installed on the triangular skylights. When the sun struck the mirrors, they took on a diamondlike brilliance, sending shafts of light throughout the space. As a result, the environment in the space continually changed. Here, images were less important than shafts of light from the mirrors (Fig. 6-20, Plate 15).

Fig. 6-18. Kevin Roche, John Dinkeloo and Associates, Richardson-Merrell Inc. Headquarters, Wilton, Conn., 1974. Detail of canopy.

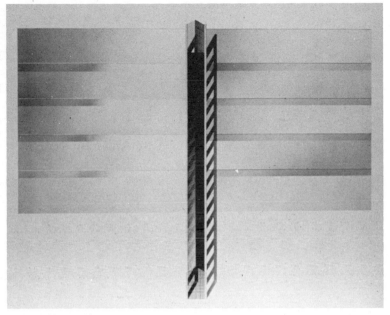

Fig. 6-19. Cesar Pelli and Associates: Biennale House project, Venice, 1976. Study used clear glass which alternated between subtle blendings of color or reflectivity.

Fig. 6-20. Cesar Pelli, for Daniel, Mann, Johnson and Mendenhall; California Jewelry Mart, Los Angeles, 1968. A combination of regular mirrors and painted panels.

Arthur Erickson's 1970 Pavilion at Osaka consisted of three triangular structures, covered on the outside with regular mirror, not semitransparent mirror. According to Erickson's perception, buildings in Japan were more earth-oriented whereas those in Canada were more sky-oriented. With the 45-degree mirrored roofs of the pavilions, he wanted to reflect images of sky and clouds into the viewer's eyes, as a means of symbolizing the vastness and sky-dominated aspect of the Canadian experience. The cross reflections from building to building, and the reflections of central mirrored "spinners," conveyed a sense of ambiguity and infinity, which Erickson felt related to the Japanese artistic attitude more than the Western one. Obviously, for Erickson, the mirrors contained considerable symbolic content. Unfortunately, an irritating side effect of those mirrors was that the ambient temperature in the immediate area was quite high due to the reflections of the sunlight. The architects ultimately had to spray the mirrors with a coating to make them less reflective (Fig. 6-21, Plate 14).

Blue Cross-Blue Shield, an office building by Odell Associates, is roughly the same size as Bell Labs, yet it is light years away in terms of overall appearance, an extreme example of "shaped glass." According to the designers, "The general appearance of the building is to be that of a simple sculptural rhomboid highly reflective to present a changing visual pattern to the passerby." A curious thing about the Blue Cross-Blue Shield building is that it looks exceedingly lightweight, more like a new form of conveyance than an actual building filled with people, desks, floor slabs and supports. It looks like it has been temporarily moored, and will soon float away into the clouds it continually reflects (Fig. 6-22).

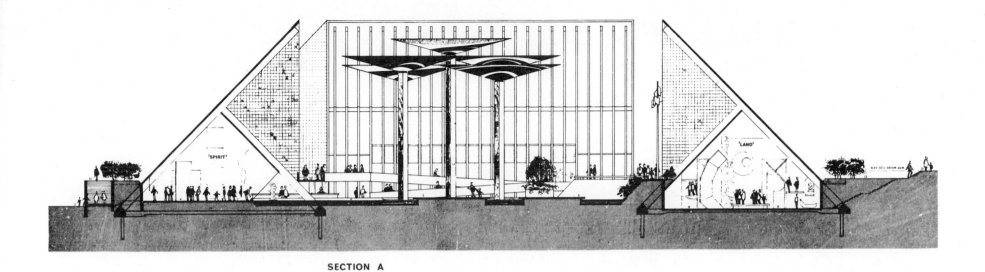

SECTION A

Fig. 6-21. Arthur Erickson and Associates: Canadian Pavilion, Osaka, 1970. Mirror cladding to symbolize the vastness and sky orientation of the Canadian experience.

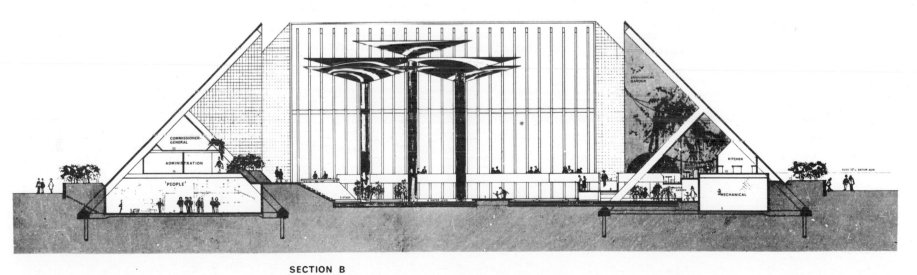

SECTION B

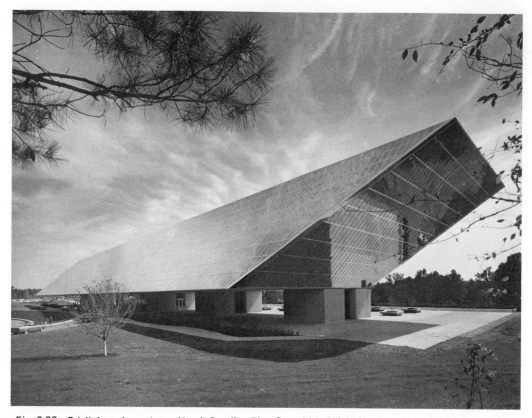

Fig. 6-22. Odell Associates, Inc.: North Carolina Blue Cross-Blue Shield Headquarters, Chapel Hill, North Carolina, 1973. This building looks like a shiny spaceship, about ready to hover away. Certainly the ultimate in dematerialization of mass.

Indiana Bell, by Caudill, Rowlett and Scott, is another example of a building for which reflections are important. Interestingly, those reflections have been controlled to a certain extent by the architects. An enormous metal vertical trellis flanks two sides of the building. The complex patterning in the trellis and the luxuriant plant growth insure a richness in the building's reflections. It also screens the surrounding urban area from the shininess of the building itself (Figs. 6-23 and 6-24, Plates 21 and 22).

John Portman has constructed numerous mirrored hotel complexes. Though they are often typified by a certain rigidity in planning, they are usually successful in the use of mirror glass. Portman defines the scale of the mirrored buildings with definite mullions, lifts them off the street from some sort of concrete podium, and juxtaposes the mirror with other buildings. For instance, Portman's Los Angeles Bonaventure Hotel is a bifurcated structure. The lower elements relate to the street. The upper towers obtain visual interest through *glaces à répétitions* (Fig. 6-25).

A particularly exuberant mirrored building is the Reunion Center in Dallas. By Welton Becket Associates, it is a large hotel complex built in a redevelopment area, just across the railroad tracks from the center of downtown. The hotel and its restaurant tower form a gateway to Dallas. Though the building is large, its mass has been fragmented with a

Fig. 6-23. Caudill, Rowlett and Scott, Inc.; Paul Kennon, design principal: Shropshire, Boots, Reid and Associates, associated architects. Indiana Bell Telephone Switching Center, Columbus, Indiana, 1978. The monumentally scaled trellis insured a richness of reflections. Photo Balthazar Korab.

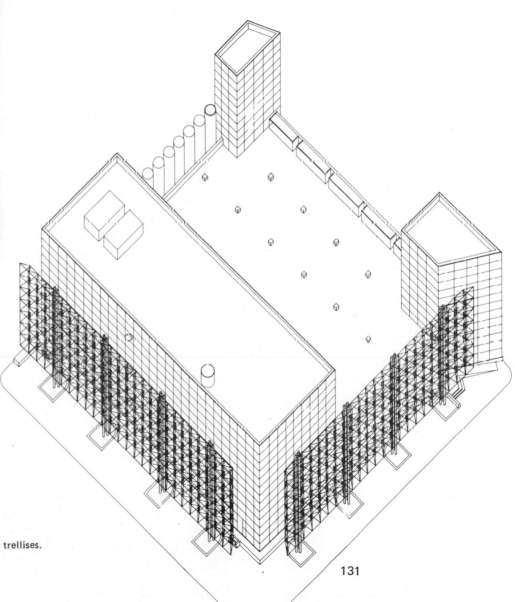

Fig. 6-24. Indiana Bell Telephone. Axonometric projection showing location of trellises.

Fig. 6-25. John Portman and Associates: Los Angeles Bonaventure Hotel, Los Angeles. The shiny towers are isolated from street level. Photo Alexandre Georges.

ziggurat shape, *glaces à répétitions*, and curving ends (Fig. 6-26, Plates 16, 17, and 18). From the scale of the automobile, it is a welcoming building (Fig. 6-27). One simply wants to go inside and see what is in that lively, shiny envelope. Yet, from the scale of the pedestrian, it suffers the problem of so many mirrored buildings (Figs. 6-28 and 6-29). One has no idea where the various floor levels are, due to the uniformity of the zipper gasket window system, and one doesn't know if there are people behind those windows. When this writer viewed the building, it was a bleak, early March day. The big Texas sky was not cheerful that day, and the building that had seemed exciting from the car seemed forbidding up close.

Two other baroquely curving uses of mirror glass were designed by Daniel, Mann, Johnson and Mendenhall. The simplest one was the Beverly Hills Hotel, Beverly Hills, California. It used curving mirrored forms to cover convention areas and shops. At their Lugano Convention Center in Switzerland, the curving mirrored forms comprise the entire building (Figs. 6-30 and 6-31).

These totally mirrored buildings are extremely difficult to handle well. Those that have been successful have been typified by restraint. Either the glass has a low degree of reflectivity, making it barely different from, say, a tinted glass building, or, if bright, the glass is treated as an accent in a given environment. The Blue Cross-Blue Shield building, for instance, is a brightly prismatic mirror, but it is remote from other structures and it reflects clouds and sky. It would be far less effective in an urban setting. In the country, it is an accent. In the city, it would be highly intrusive.

Fig. 6-26. Welton Becket Associates: Reunion Center, Dallas, 1978. A baroque use of *glaces à répétitions*. Photo Balthazar Korab.

Fig. 6-27. Welton Becket Associates: Reunion Center, Dallas, 1978. A gateway building. Photo Balthazar Korab.

Fig. 6-28. Welton Becket Associates: Reunion Center, Dallas, 1978. Entrance to building. Photo Balthazar Korab.

Fig. 6-29. Welton Becket Associates: Reunion Center, Dallas, 1978. Photo Balthazar Korab.

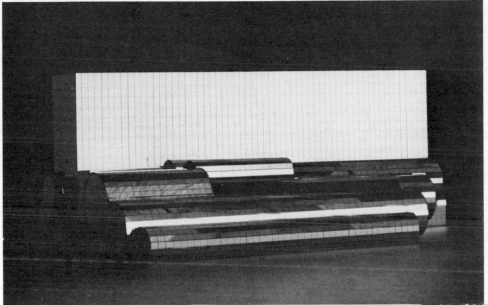

Fig. 6-30. Daniel, Mann, Johnson and Mendenhall: Beverly Hills Plaza, project Beverly Hills, California, an exuberant example of shaped glass. Photo Lang.

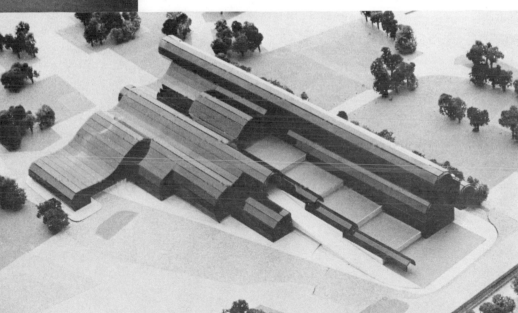

Fig. 6-31. Daniel, Mann, Johnson and Mendenhall: Lugano Convention Center project, Lugano, Switzerland.

135

7
Contrasts and Complements

Mirror glass buildings that are total glass expressions frequently suffer from scale problems. When the glass is combined with other materials, however, such as metal spandrels or brick walls, the scale problem is often lessened, or even eliminated. The other materials usually help to define function and size. So the other materials "help" the mirror glass. The glass, on the other hand, sometimes affects the other materials in a positive way. A massive building might seem less heavy and ponderous with mirrors as accents.

Roche/Dinkeloo have for some time combined mirror glass with concrete walls as a form of contrast. Roche has said, "The glass is quite extraordinary in terms of lessening the weight of the wall." Whereas the French lightened visually the interiors of their *châteaux* with mirrors, modern designers have used the material on the outside to make the walls seem less massive. For Roche, the glass is often a distinct contrast to the other materials.

If the other materials are lightweight looking, such as a shiny aluminum spandrel, the mirror glass might become a complement to it. The total effect is a taut expression of glass and metal skin, a continuation of international style attitudes, albeit in a shinier new package.

Sometimes the illusions that play on the glass are important and sometimes they are not. And since mirror glass can vary in terms of reflectivity, sometimes contrasts between the glass and other materials are subtle, sometimes they are great. John Deere, the first totally mirror glass building, actually read as a steel building with a subtle, slightly hazy glass infill. The mirror glass had a soft, gold tint to it. The reflectivity of the glass was barely noticeable, and the building had a look of permanence about it.

Roche/Dinkeloo's Power Center for the Performing Arts also had a look of permanence about it. There, the glass was used in a broader stroke as something of a mural, and was a brighter contrast to the solidity of the overall design. The glass at the Power Center duplicates the massive columns, creating an illusionary arcade. It completes the semicircular stair tower, making it a cylinder visually. And it duplicates a small yet particularly striking stand of trees. The mirror glass at Power Center is rather like decoration on one side of the building (Figs. 7-1, 7-2 and 7-3).

Roche/Dinkeloo combined mirror glass with concrete at the Cummins Engine Co. Headquarters. The glass was treated as small, bright strips, not so much to convey views of trees,

Fig. 7-1. Kevin Roche, John Dinkeloo and Associates: Power Center for the Performing Arts, U. of Michigan, Ann Arbor, 1971. The glass and the building are contrasting textures.

Fig. 7-2. Power Center for the Performing Arts. The reflection of the park is an important aspect of this design. The colonnade is also duplicated in the glass.

Fig. 7-3. Power Center for the Performing Arts. The circular form of the stair tower is completed by the mirrored wall.

but to make the concrete wall seem more lightweight. One can see similar treatment at One Summit Square, and at the John Deere Insurance Company. The latter is particularly intriguing, for mirror glass is used in a *trompe l'oeil* fashion. At a 45-degree angle, a mirror glass window wall reflects a plain wall. Hence, a glass wall is made to look like a solid material (Figs. 7-4, 7-5, 7-6 and 7-7).

At the Alfred C. Glassell, Jr. Art School, mirror glass was used in the form of a solid material. Studio spaces were clad with a reflective glass block that transmitted subtle views into the building. The texture of the block avoided the slickness so often a problem with mirror buildings. Images on the block were subtle and fragmented (Fig. 7-8).

Paolo Riani, an Italian architect who practices in the United States as well as abroad, creates a sense of richness with his mirror facades. At his MEC Design showroom in Tokyo, the glass facade was treated as an enormous sign, with the letters for the shop applied to the mirror itself. Views into the shop were permitted with the clear glass door and display windows. The effect was flashy and exciting as the windows reflected the kinetic urban presence with a diagonal window grid. The whole composition was contained with a poured in place concrete frame (Figs. 7-9 and 7-10).

Fig. 7-4. Kevin Roche, John Dinkeloo and Associates: Cummins Engine Co. Headquarters, Columbus, Indiana. The strip windows are designed to be mirror glass in order to lighten the weight of the wall visually.

Fig. 7-5. Cummins Engine Co. Headquarters.

139

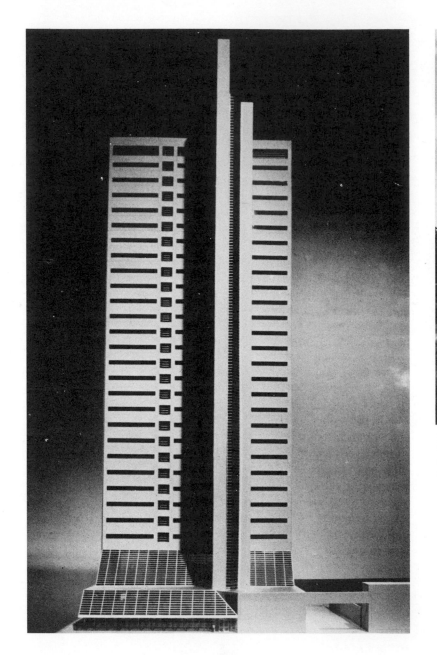

Fig. 7-7. Kevin Roche, John Dinkeloo and Associates: John Deere Insurance Co., Moline, Illinois. Mirror glass usually brightens the wall. In this case, the glass resembles it. Strip windows just left of center are at a 45 degree angle. They mirror the blank wall of a protruding element in the center.

Fig. 7-6. One Summit Square, Fort Wayne, Indiana, 1980. Mirror glass windows, skylights and canopies as a contrast to the building's mass.

Riani used a similar diagonal mullion system at his Hampden Country Club in Springfield, Massachusetts. Three sides of the building were clad simply in white metal panels. Then, the facade flanking the golf course became a celebration of nature. The glass was held in place with diagonal white metal mullions, looking rather like a monumentally scaled trellis. Then, in a particularly sensitive stroke, the glass facade was tilted at a 30-degree angle. Because of this tilt, it meant that the views of trees and grass, rather than the glare of the sky, would be reflected to the viewer. Since the building was sited on the crest of a hill, had the facade been vertical, it would have reflected a greater amount of sky to the viewers below (Figs. 7-11 and 7-12, Plates 19 and 20).

A small scaled and pleasant use of mirror glass was achieved by Warren Platner at the Teknor Apex Headquarters. It features an arcaded courtyard, flanked on one side by a mirror glass curtain wall. Due to the low height of the glass, one is primarily confronted with reflections of trees rather than sky (Fig. 7-13).

Residential applications of mirror glass have been rare. One of the first designers to use the material in a private residence was James Lambeth. In the subsequent section, one will see how he used mirror glass in a solar fashion. In two early residences, he was concerned more with images on the glass itself rather than redirecting the sun's rays. The two residences shown are similar in that they are both bifurcated and accessed by bridges. Lambeth's own residence, with bedrooms behind the mirror glass, has walls angling toward the bridge. The walls provide privacy for the bedrooms and mirror the bridge, surrounding trees, as well as the walls themselves. The other residence, the McKamey residence, had the walls sloping upward toward the sky (Figs. 7-14 and 7-15).

Fig. 7-8. S. I. Morris; Eugene Aubrey, designer: Alfred C. Glassell, Jr. Art School, Houston, 1979. Glass block with a reflective coating. (*Courtesy Pittsburgh Corning*)

Fig. 7-9. Paolo Riani Associates: showroom of MEC Design International, Tokyo, 1971. The glass fragments the exterior view.

Fig. 7-10. Showroom of MEC Design International.

Fig. 7-11. Paolo Riani Associates: Hampden Country Club, Springfield, Massachusetts, 1974. Since the glass is tilted downward, it reflects trees and lawn more than sky.

Fig. 7-12. Hampden Country Club.

Fig. 7-13. Warren Platner Associates: Teknor Apex Headquarters, Pawtucket, Rhode Island, 1974. A restful garden is duplicated by the glass. Photo Ezra Stoller.

Fig. 7-14. James Lambeth: residence, Fayetteville, Arkansas, 1969. An unusual use of mirror glass in a residence.

Fig. 7-15. James Lambeth: McKamey residence, Fayetteville, Arkansas, 1974. Images are important here.

144

At Gunnar Birkerts' Federal Reserve Building in Minneapolis, the mirror glass is a lightweight infill, in contrast to the monumental pylons at the ends of the building. The pylons were installed in order to avoid columns in the vault areas below. Said Birkerts of the design, "The desirable large span and the subsequent unique structural system were used to create the bold expression responding symbolically to the majestic scale and vastness of the landscape of the Northwest." According to him, the glass below the arch was brought flush with the structure as a means of protecting it. As in many of the combination buildings, the glass at the Federal Reserve does not conceal the structure, but helps make it all the more visible (Figs. 7-16 and 7-17).

COMPLEMENTS

At the Federal Reserve, the glass is a contrast to the massive masonry pylons supporting it. In many other combined forms, the glass is not so much a contrast to the other materials, but a complement, or a subtle continuation.

Birkerts' IBM Computer Center is a glass and metal panel curtain wall building in which the entire structure looks lightweight. The glass alone does not look lightweight, but the metal panels do as well (Figs. 7-18 and 7-19).

The client wanted inexpensive space with little expression of corporate image since the site was isolated. According to Birkerts, in *Architectural Record*:

A glass and metal curtain wall is still the cheapest way to enclose space today, on any project larger than a house. A reflective surface does reduce the internal energy required to run a building (whatever else it may do). My choice of these two elements—curtain wall and reflective surface—for Sterling Forest, plus the use of the simplest form, a cube, were for me absolutely pragmatic . . .

Fig. 7-16. Gunnar Birkerts and Associates: Federal Reserve Building, Minneapolis, 1973. Images on the glass are less important than an expression of lightweight, taut glass skin. Photo Balthazar Korab.

145

Fig. 7-17. Gunnar Birkerts and Associates: Federal Reserve Building, Minneapolis, 1973. Photo Balthazar Korab.

As a means of symbolizing internal functioning of the building, Birkerts clad the office areas with reflective glass. The spaces housing machines were enclosed with solid reflective aluminum panels. A red metal "stripe" was inserted between the two areas, a symbolic danger zone, implying that the machines, which were closed and significant, should not overwhelm or diminish the people. Birkerts further described the glass itself:

It is a symbol of a particular kind of abstract intellectual work, and of the computer itself. It seemed necessary to contrast all this as sharply as I could with the setting, with nature. The mirror surface is about technology today; it is not postulated as a projection of future conditions, for we may turn against this technology.

Sterling Forest is an exceedingly precisely detailed building. With round ventilation holes, ship railings and expressions of skin, it is a technological expression. However, the irony of Sterling Forest is that due to the perfection of its skin, it reflects nature with a clarity that almost makes it merge visually into the setting. Due to its reflections, it almost looks romantic (Fig. 7-20).

Trees have a way of making mirrored buildings look romantic; sky has a way of making them look cooly technical. The Federal Aviation Administration Building by Daniel, Mann, Johnson and Mendenhall reflects no trees, just sky and surrounding suburban landscape. As a result, its technical expression is more convincing. The edges of the form are defined with smoothly curved metal panels, themselves almost expressive of curving airplane forms (Figs. 7-21 and 7-22).

Fig. 7-18. Gunnar Birkerts and Associates: IBM Computer Center, Sterling Forest, New York, 1972. The machined cube in the landscape. At this level, the building is a contrast with its environment. Photo Balthazar Korab.

146

Fig. 7-20. Gunnar Birkerts and Associates: IBM Computer Center, Sterling Forest, New York, 1978. In this view, the building is no longer in complete contrast with nature. Photo Balthazar Korab.

Fig. 7-19. Gunnar Birkerts and Associates: IBM Computer Center, Sterling Forest, New York, 1978. Building's function defined, not concealed. Computers behind metal wall, offices behind glass wall. Photo Balthazar Korab.

Fig. 7-21. Daniel, Mann, Johnson and Mendenhall: Federal Aviation Administration Building, Hawthorne, California, 1975. A precise technological expression. Photo Wayne Thom.

147

Fig. 7-22. Daniel, Mann, Johnson and Mendenhall: Federal Aviation Administration Building, Hawthorne, California, 1975. Photo Wayne Thom.

Johnson-Burgee's Post Oak Central is also a curving expression of skin. However, at Post Oak, the glass is applied in smaller ribbon windows. Particularly in photographs, this type of building doesn't read as a mirror glass building. It has no scale problems and is not overly abstract due to the horizontal spandrels. Reflections on the facade are discontinuous, also due to the spandrels. It looks sleek and streamlined (Fig. 7-23).

Streamlining as a concept developed in the 1930's as cars, trains and planes got rid of their previously boxlike appearances and became sleek, ovoid packages so as to move more easily through the wind (Fig. 7-24). The sense of movement was often emphasized by industrial designers with the reduction of the size of vertical window supports, and the application of horizontal lines or stripes on the exterior of the package. Even toasters and washing machines had little lines applied to them. And buildings developed a sleek, packaged look with horizontal ribbon windows. Siegfried Giedion had noted Americans' love of horizontality even in the nineteenth century. In *Space, Time and Architecture*, he pointed out the prevalence of horizontal-looking porches on American houses vis-à-vis those of their European counterparts. He said, "Apparently, Americans like a strong and unbroken horizontal line equally in their houses and in the Pullman car."

Cesar Pelli used horizontal ribbon windows and spandrels of concrete at the U.S. Embassy building in Tokyo (Plate 23). The State Department had specified that the building

Fig. 7-23. Philip Johnson and John Burgee: Post Oak Central, Houston, 1980. A streamlined building. Photo Richard W. Payne.

Fig. 7-24. Amtrak's Broadway Limited. The illusion of rapid horizontal motion is intensified by a horizontal banding of the windows. (*Courtesy Amtrak*)

Fig. 7-25. The Architects' Collaborative: Johns-Manville World Headquarters Building, Denver, 1977. A combination of lightweight looking materials, horizontally disposed. Photo Nick Wheeler.

be concrete. He treated it as a "skin," having little difference between the face of the concrete and the glass itself. The concrete was painted with a glossy paint to intensify the lightweight look that Pelli wanted. Though Pelli rarely uses reflective glass in his work, preferring the subtlety of tinted glass, he inserted reflective glass ribbon windows at the Embassy in order to make the concrete "more lively." Also, being an embassy structure, he felt that it would be appropriate if people from the outside could not see in.

Some of the streamlined buildings seem expecially light-weight. One such is the Citicorp building by Hugh Stubbins, in Manhattan (Plate 24). The building has a daring structural system. The four corners of the building are canti-levered out over the site to allow for pedestrian concourses and a church below. Yet, those four corners of the building carry a considerable amount of load from the stories above. What makes it seem not too alarming is partly due to the facade treatment. One is aware of shiny horizontal bands on the building, comprised of aluminum spandrels or bright silvery windows. No columns are visible behind the shiny facade. Hence, one is not aware of vertical loads plum-meting to the four corners, but one notices the appearance of extraordinary lightness. The building looks like a series of shiny, weightless trays placed on top of one another.

At Johns-Manville World Headquarters, by the Architects Collaborative, the building is as light and streamlined looking as a train (Plates 25 and 26). It looks as if it might glide away any instant. It is a decidedly horizontal structure, 1000 feet long, poised above the foothills on a 10,000-acre ranch outside Denver. Though this building is clad in a fairly bright reflective glass (33 percent), many architects do not realize that it is a mirror glass building. It just looks particularly sleek and modern, a technological expression that is a contrast with its surroundings, yet one that does not compete with nor overwhelm its environment. When looking at Johns-Manville from the front, one is completely unaware that cars are parked in and behind the structure (Figs. 7-25, 7-26 and 7-27).

The interior mirror might appear to be a glassy plane separating one space from another, or pure space itself. The mirror outdoors is less illusionary. Yet, at times, such as in the case of Johns-Manville, one might feel one is looking at a clear glass building—albeit with a particularly clean and sparkling presence (Figs. 7-28 and 7-29).

Fig. 7-26. Johns-Manville World Headquarters Building. Images are less important than a crisp, machined expression. Photo Nick Wheeler.

Fig. 7-27. Johns-Manville World Headquarters Building. Photo Nick Wheeler.

151

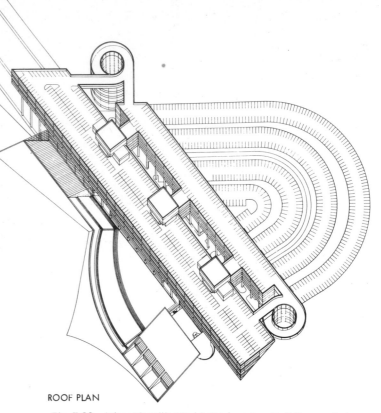

ROOF PLAN

Fig. 7-29. Johns-Manville World Headquarters Building, roof view.

Fig. 7-28. Johns-Manville World Headquarters Building. Photo Nick Wheeler.

152

8
The Solar Mirror

*The concave mirror although cold when it receives the
rays of the fire, reflects them hotter than the fire.*

Leonardo da Vinci

In the summer of 1978 James Lambeth, an architecture professor at the University of Arkansas, organized some students and friends into recreating the myth that Archimedes caused a Roman fleet to burn, all with the aid of mirrors. The group was provided with hand-held mirrors to be focused at small scale models of Roman boats. It was intended to be a celebratory gesture for Sun Day, a national day set aside to instruct and introduce Americans to the potential of solar energy. NBC News was waiting with reporters and minicams to film the conflagration. The only problem was the sun. It did not shine in Arkansas on Sun Day, nor for the next two days. Fortunately for the state of Arkansas, the Roman fleet was only made of cardboard. When Lambeth was asked if the tale of Archimedes had any basis in fact, he said it was "definitely legend" (Fig. 8-1).

Whether the tale of Archimedes is true or not, the fact remains that in the United States since the early 1970's, solar mirrors have performed a number of functions with reliability. Some architects have created customized mirrors while others have utilized off-the-shelf, pre-engineered products.

James Lambeth had been using mirrors in his work since the late 1960's. Initially, he had been involved with psychedelic mirrored experimentation. Then he constructed two houses with large amounts of mirror glass, his own and the McKamey residence. He realized that with a subtle shifting of the reflective surfaces, he could turn the mirror from merely an aesthetic object to an aesthetically pleasing tool. He wrote: "It became apparent the images were only a part of the phenomena of the reflective surface. The paths of reflected sunlight that fell around those designs led to new directions." He then described how he began to conceive of his reflective entrance at a ski lodge, Yocum Lodge, built in the early 1970's. He wrote, "My memories of reflected sun began to emerge as a focusing lens above the vulnerable entry. With proper arrangement, the lens would melt the snow and ice throughout the ski season."

Yocum Lodge is fascinating primarily because of the way it used mirror glass. A 90 percent reflective mirror glass window was installed on the second floor of the dwelling, just above the doorway. Instead of being vertical, the usual way of using the material, the individual panes of glass were installed to form a subtle arc. The angles of the individual panes were calculated to reflect low-in-the-sky winter sun onto the entry area, to eliminate the burdensome task of shoveling snow. In the summer, when the sun was higher in

153

Fig. 8-1. Recreation of the myth of Archimedes. Organized by Prof. James Lambeth with the U. of Arkansas, Fayetteville, Arkansas, 1978.

the sky, the glass was essentially in its own shadow. Residents in surrounding ski houses were somewhat nervous about the concentrator, fearing the possibility of some horrible accident. But according to Lambeth, it was a "gentle" lens. The solar reflections were directed in a linear fashion rather than to a specific point (Fig. 8-2, Plate 30).

The Lambeth house had nothing to do with solar heating inside the dwelling. It was simply an ingenious solution to a task that has bothered man for centuries. It also used mirror glass in a dynamic fashion so that the reflected solar energy was concentrated instead of being dispersed (Figs. 8-3 and 8-4).

A vacation house designed shortly after the Lambeth house used a movable system of reflectors. Created by Michael Jantzen, it used solid shutters with a reflective coating on them to reflect sunlight through a skylight down to a water storage tank. The shutters were operated with a simple winch. They could be adjusted for varying solar angles, and could be closed at night. Since the shutters were insulated, they helped to retain heat in the water storage tank. According to Jantzen, the motivation for the house was not the energy crisis. Said he, "It was done in response to a personal need to design an energy conscious living machine. Energy crisis or no energy crisis, I would have done it. The house works very well" (Figs. 8-5 and 8-6, Plates 28 and 29).

From an aesthetic standpoint, the Lambeth ski lodge and Jantzen vacation house were quite different. The Lambeth house was picturesquely designed, using the local vernacular of wood siding and pitched roofs. The mirrored lens was a distinct contrast, in terms of form and in terms of material, to the rest of the design. In the Jantzen house, the mirrored shutters represented a continuation of the high tech imagery of the rest of the design. They were another crisp, metallic surface.

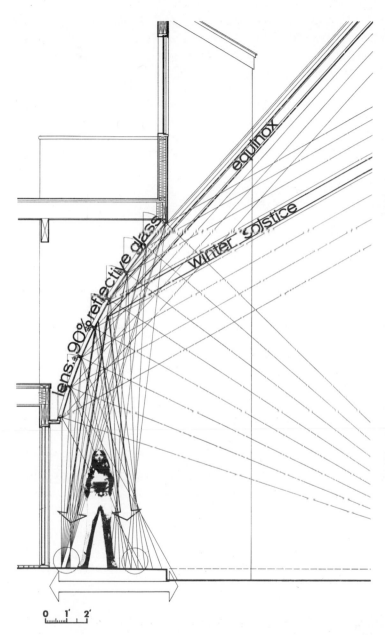

lens: 90% reflective glass

equinox

winter solstice

0 1' 2'

Fig. 8-3. Yocum Lodge, winter view. In summer, mirrors are in their own shade.

Fig. 8-2. James Lambeth: Yocum Lodge, Snowmass at Aspen, Colorado, 1972. Solar diagram. Note the direction of the reflected rays of sun emerging from the mirror glass windows.

155

Lambeth continued to experiment with a wide variety of reflectors. Some were mirror glass, some were movable flaps, some were solid lenses. In two swimming pool designs, the pools are covered with glass greenhouselike structures. Hovering over those greenhouses are canted 90 percent reflective mirror glass. The pools were to be warmed with direct and reflected radiation. An apartment building called "Solar Pac" was also designed with 90 percent reflective mirror glass windows (Fig. 8-7). They were canted so that the energy would strike solar collectors at the windowsill. A "Fontainbleau Retrofit" scheme theorized that the windows on the old hotel could be revised to act as solar reflectors. Then in a grandiose scale, Lambeth designed an "Alpine Lens" that would hover over a mountaintop greenhouse of immense size, showing the possibilities of direct and reflected radiation in remote, cold regions.

Even the lowly mobile home received his attention. He designed a prototype to be oriented with the long wall facing south. Foldout flaps at the top and bottom would be awning and terrace, respectively. Foldout flaps at the sides would be solid mirrored panels. Their purpose would be to increase wintertime radiation.

Lambeth won a competition in Japan with a prototypical housing scheme. The roofs of the houses were to be trough-shaped, metal reflectors. The roofs were to be tilted at an angle of 38 degrees, perpendicular to the equinox at latitude 36 degrees N. Above the roofs were to be moving rods which would contain water to be converted into electricity and used in absorption cooling. According to Lambeth, the heat generated was to have been 1100 degrees Fahrenheit. Although this is higher than usually considered appropriate for residential design, the system obviously could be modified to conform to realistic requirements. From an aesthetic standpoint, the project has immense appeal. Lambeth thought of the project as "a regatta of silver sails on a crystal lake" (Figs. 8-8 and 8-9).

Fig. 8-4. Yocum Lodge, detail of reflector.

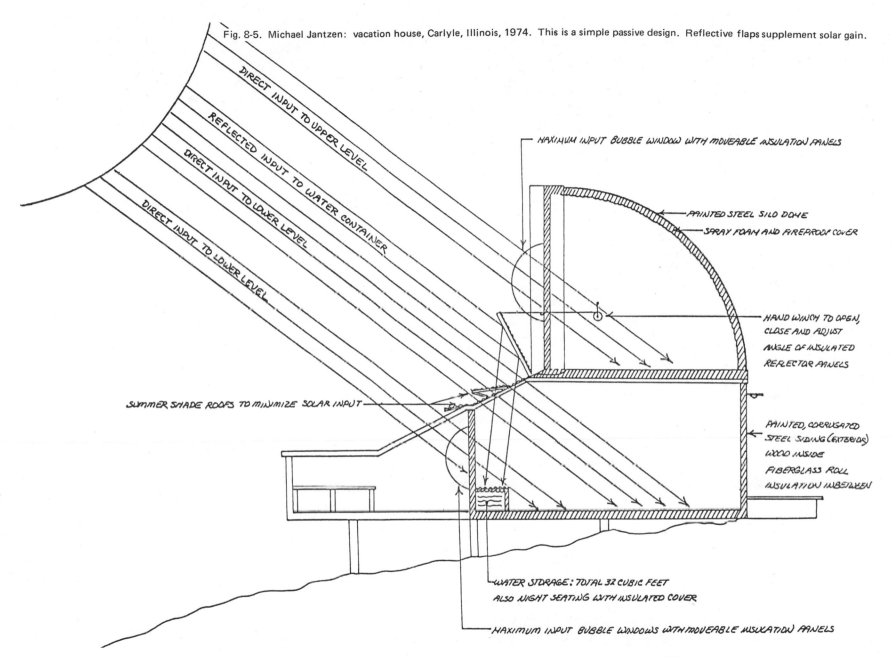

Fig. 8-5. Michael Jantzen: vacation house, Carlyle, Illinois, 1974. This is a simple passive design. Reflective flaps supplement solar gain.

DIRECT INPUT TO UPPER LEVEL

REFLECTED INPUT TO WATER CONTAINER

DIRECT INPUT TO LOWER LEVEL

DIRECT INPUT TO LOWER LEVEL

MAXIMUM INPUT BUBBLE WINDOW WITH MOVEABLE INSULATION PANELS

PAINTED STEEL SILO DOME

SPRAY FOAM AND FIREPROOF COVER

HAND WINCH TO OPEN, CLOSE AND ADJUST ANGLE OF INSULATED REFLECTOR PANELS

SUMMER SHADE ROOFS TO MINIMIZE SOLAR INPUT

PAINTED, CORRUGATED STEEL SIDING (EXTERIOR) WOOD INSIDE FIBERGLASS ROLL INSULATION INBETWEEN

WATER STORAGE: TOTAL 32 CUBIC FEET ALSO NIGHT SEATING WITH INSULATED COVER

MAXIMUM INPUT BUBBLE WINDOWS WITH MOVEABLE INSULATION PANELS

157

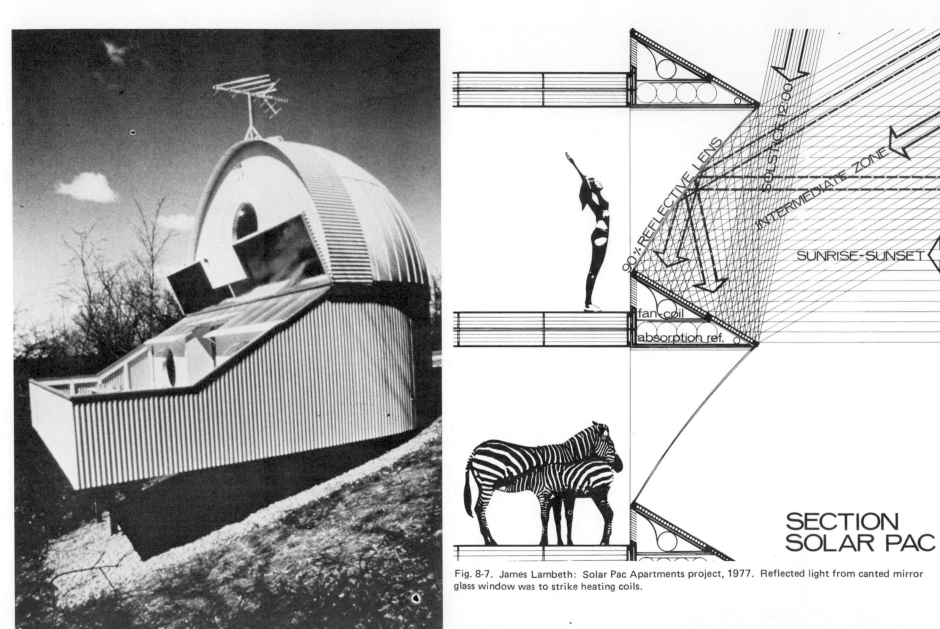

SECTION
SOLAR PAC

Fig. 8-7. James Lambeth: Solar Pac Apartments project, 1977. Reflected light from canted mirror glass window was to strike heating coils.

Fig. 8-6. Jantzen vacation house. The reflective flaps are easily adjusted.

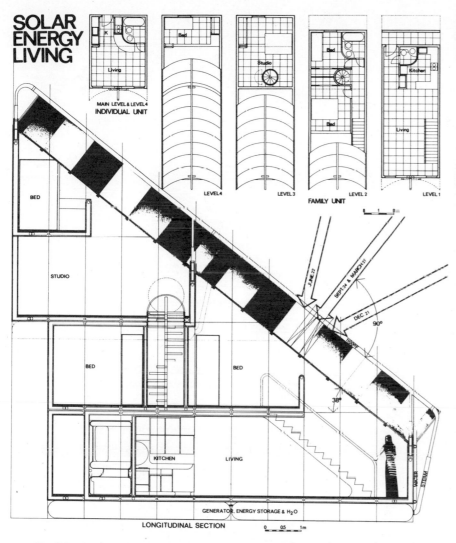

SOLAR
ENERGY
LIVING

MAIN LEVEL & LEVEL 4
INDIVIDUAL UNIT

LEVEL 4   LEVEL 3   LEVEL 2   LEVEL 1

FAMILY UNIT

BED

STUDIO

JUNE 22

SEPT 24 & MARCH 21

DEC 21
90°

BED   BED

38°

KITCHEN   LIVING

WATER
STEM

GENERATOR, ENERGY STORAGE & H₂O

LONGITUDINAL SECTION   0   0.5   1m

Fig. 8-8. James Lambeth: Misawa Homes project, 1975. The form of this roof
derives from engineering parabolic reflectors with moving collector rods.

Fig. 8-9. James Lambeth: Misawa Homes project.

159

Fig. 8-10. James Lambeth: Strawberry Fields Apartments, Springifeld, Missouri, 1976. The custom-made lens heats water in the pool during spring and fall months.

To date, the Misawa Homes project, the mobile home, and the various mirror glass lens studies have not been built. They exist as architectural ideas, though not at this time as reality.

A large-scale lens to be realized was part of the Strawberry Fields apartment complex. It was custom-made stainless steel, designed to heat water in a solar collector so that the apartment swimming pool could be used in spring and fall months. This extended the swimming season by warming the water 10 degrees. The lens was angled so that the collector would be in shade in the summer when the sun was high, and when solar heat was not needed. It was operative only in the cool months when it was needed (Figs. 8-10 and 8-11, Plate 31).

In Lambeth's continued experimentation with reflection, he located a metallized plastic that would fragment the rays of sunlight that struck it. The rays of sun would emerge as rainbows, thereby imparting a decorative hue to any object in their path. His idea was to install the material on a low roof adjacent to a south-facing wall. When the light struck the low roof, it would reflect upward to the other wall. It would thereby impart additional reflected heat to that wall, as well as varying shades of yellow, red and blue. While this application is strictly speaking not a mirror, it is an intriguing development derived from Lambeth's earlier experimentation with reflection (Fig. 8-12).

The end result of much of Lambeth's solar work is simply to make life more pleasant for people. The schemes go beyond mere comfort in dwellings. People shouldn't have to trouble themselves with shoveling snow since the reflector can do that. If they want to swim for more than just three months a year, the reflector can help them do that as well. And why not give them rainbows on a facade when using reflected solar energy? As a result, one gets a genial sense of humanity in Lambeth's work.

A conceptual work by a different architect, Frederick Fisher, was designed not as a celebration of life but as an ode to death. His project was a solar crematorium. It

160

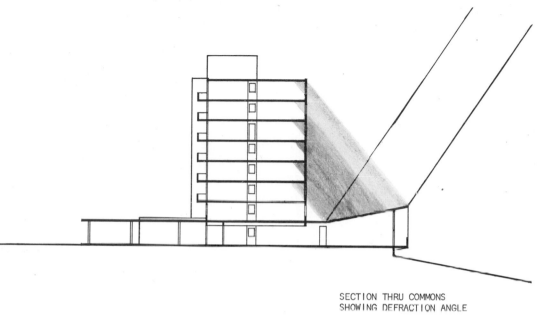

SECTION THRU COMMONS
SHOWING DEFRACTION ANGLE

Fig. 8-12. James Lambeth: Housing and Urban Development housing project: Gorman Towers, Fort Smith, Arkansas, 1976. With an embossed aluminum metallized polyester film on the low roof, light will not simply be reflected. It will be refracted into its constituent colors, creating a rainbow effect on the adjacent facade.

Fig. 8-11. James Lambeth: Strawberry Fields Apartments, detail of lens.

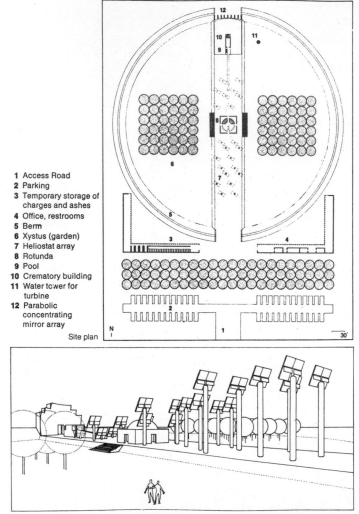

1 Access Road
2 Parking
3 Temporary storage of charges and ashes
4 Office, restrooms
5 Berm
6 Xystus (garden)
7 Heliostat array
8 Rotunda
9 Pool
10 Crematory building
11 Water tower for turbine
12 Parabolic concentrating mirror array

Site plan

Fig. 8-13. Frederick Fisher: prototypical solar crematorium. Winner of a citation in *Progressive Architecture* magazine Design Awards Program, Jan., 1977. The form is derived from Odeillo. (*Courtesy Progressive Architecture*)

bore a strong resemblance to the solar furnace at Odeillo, France. It consisted of heliostats and a parabolic concentrating mirror array to focus solar energy to a small cremation chamber. The project won a citation in the 1977 *Progressive Architecture* magazine Design Awards Program. The jury's comments were diverse. Craig Hodgetts said, "It is a poetic manifestation of a coming technology," while Sarah Harkness remarked, "This one does seem morbid [referring to Finnish crematorium designs] whereas an ordinary crematorium would not be. It brings you to thinking of things like sacrifices" (Fig. 8-13).

From the standpoint of both the living and the deceased, the project was over-designed. Once the mourners exited from their cars, they were forced by the design to walk the equivalent of two city blocks to the centrally-sited rotunda for the initial ceremony. The casket would then be borne almost another block to the crematorium. When a solar scientist reviewed this project, he said that such an elaborate array of heliostats and focusing mirrors was unnecessary for the intended use. According to him, one parabolic dish, 20' in diameter, would introduce 2000 degrees on the object. That would be the equivalent of the heat of 300 suns in one spot. As he phrased it, "That could give you a good sunburn."

A somewhat less grandiose conception, but with fascinating implications, was devised by architects Banwell, White and Arnold for the Famolare shoe factory in Vermont. The architects wanted more sun to enter a building through clerestory windows. They had mirrors mounted on the roof. The mirrors were operated by a pivot to account for different solar angles. Not tracking mirrors, they were merely seasonally adjusted to reflect sunlight into the building in the winter and away from the building in the summer. The fascinating thing about these mirrors was that sun would now enter the building from the north, through the power of reflection (Figs. 8-14, 8-15 and 8-16).

162

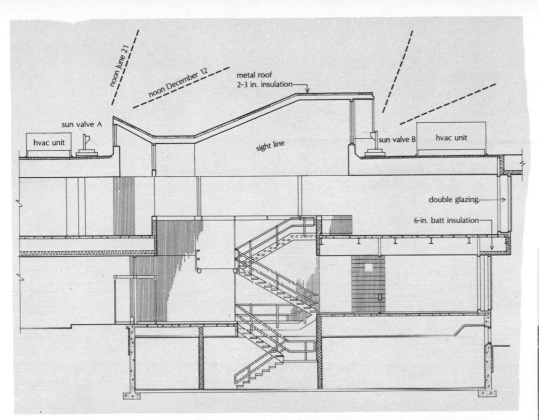

noon June 21
noon December 12
metal roof 2-3 in. insulation
sun valve A
hvac unit
sight line
sun valve B
hvac unit
double glazing
6-in. batt insulation

Fig. 8-14. Banwell, White and Arnold: headquarters for Famolare, Inc., Brattleboro, Vermont, 1978. Movable mirrored panels, called "sun valves" direct sunlight to the center of the building.

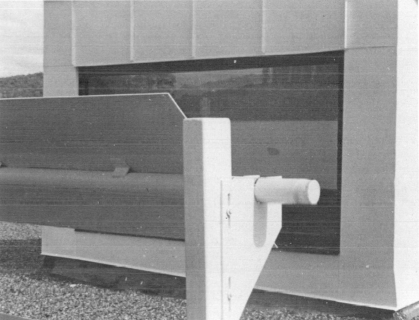

Fig. 8-16. Famolare sun valve at north side of building.

Fig. 8-15. Famolare sun valve at south side of building.

Fig. 8-17. The Architects Taos; William Mingenbach, Harold R. Benson, principals in charge: Mary Medina Building, Taos, New Mexico, 1978. (*Courtesy Koppers Roofing*)

Another case of sunlight entering a building through clerestory windows occurred at the Mary Medina building in Taos, New Mexico, by a design group called the Architects Taos. A highly spectral aluminum membrane roof was installed on the building. Since the roof was serrated, light would bounce off the roof and strike large water drums placed behind clerestory glazing at the roof level. Reflective materials were placed on the ceiling above the drums, both to intensify energy gain further and to reflect light into the center of the building. This was a desirable effect since the building had few exterior windows due to a desire to conserve energy. Another energy conserving measure were insulating shutters to cover the clerestory windows from the inside as a means of insulating them at night. The system was designed to meet 60 percent of the structure's heating requirements and 70 percent of its lighting requirements (Figs. 8-17 and 8-18).

Most of the architectural projects described have featured custom-designed reflectors, sending solar energy downward, upward or sideways, depending on where the architects wanted the extra energy. In many cases, the building forms responded to the dynamic way of using the mirror. Lambeth's angled walls, Jantzen's movable shutters, the serrated roof at the Mary Medina building: all reached for the sun in a new way.

Higher operating temperatures are likely with prefabricated solar collectors, for they are calculated to receive maximum sunlight throughout the day, and often track the sun. As for affecting the look of architecture, in certain instances, they do. But often the collectors are installed in banks remote from the building itself.

Fig. 8-18. Mary Medina Building, detail of roof. A highly spectral aluminum membrane roof was used to reflect light to water drums located just behind the clerestory windows. (*Courtesy Koppers Roofing*)

The compound parabolic concentrator developed by Dr. Roland Winston, a physicist at the Argonne National Laboratory, has the ability to gather both direct and diffuse solar radiation without tracking the sun. It can work on a cloudy day as well as a sunny day, creates temperatures from 140 to 400 degrees, and has a low profile. The total depth of one of these units might be six inches, including mirrors and concentrating tubes. The lack of bulk of these collectors makes it fairly simple to integrate them into the facade of a building. Since they look so much like standard flat plate collectors, one is not particularly aware that one is looking at a mirrored installation (Figs. 8-19 and 8-20).

Larger scale parabolic solar collectors are installed at Honeywell Corporate Headquarters complex in Minneapolis. The field of 252 collectors is located atop an employee parking garage, and is designed to provide energy for a new eight-story office building. On sunny days, the system is designed to provide all the energy needed to heat and cool

Fig. 8-20. Installation showing compound parabolic concentrator. (*Courtesy Solara Corp.*)

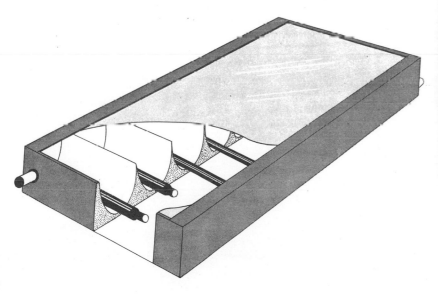

Fig. 8-19. Compound parabolic concentrator. The mirrors are metallized and laminated with polyester. This collector works with diffuse as well as direct radiation. (*Courtesy Solara Corp.*)

165

Fig. 8-21. Hammel Green Abrahamson, Inc.: Honeywell Plaza, Minneapolis, 1978. Solar collectors installed atop a parking garage.

for cooling was met by the system, reducing the electrical needs of the building during peak demand hours. At both Georgia Power and Honeywell, the look of the buildings was not appreciably affected by the presence of the mirror field. However, from upper floors, they are probably more pleasant objects to look at than tar and gravel roofs or parked cars on a garage roof (Fig. 8-24).

Some of the large-scale, high-temperature solar mirrors are intriguing objects from a strictly visual standpoint. The shapes are bold, reflections blue and bright. And there is a sense of surprise that a material we associate with living rooms can act in such a powerful fashion. They are not works of architecture, yet are often the scale and have the visual appeal that works of architecture do. As architects in the past have responded to the visual characteristics of steamships, airplanes, grain silos or suspension bridges, so they are likely to respond to the beauty of these examples of a technology that is becoming more prevalent.

A proposed power plant for Crosbyton, Texas, pop. 2500, is likely to be one of the most beautiful in the world. In the midst of fields on the outskirts of town will sit ten 200' in diameter mirrored bowls, like giant cereal bowls, set in an even row. Funded by the Government, they are designed to provide 30 percent of the electric energy for the town, with the remainder provided by fossil fuels. They are fixed mirror systems. Though the bowls are tilted to the south, the tracking is done by boilers suspended overhead. This is different from the power tower concept, where the tower containing the boiler is stationary and the mirrors do the tracking. At Crosbyton, the mirrors will create steam at 1000 degrees Fahrenheit for a five-megawatt electric generating system (Fig. 8-25).

Given our poor storage capabilities of electricity, mirror systems and other non-fossil fuel generating systems will need an alternate energy backup. Yet, the mirror system just might play an increasingly visible role in the future as one form of solar energy, an energy source that won't have to be hidden at the far edge of the sight.

the building. In the course of a year, over 50 percent of the building's heating requirements, more than 80 percent of the cooling and all of the hot water needs will be met by the field of tracking mirrors (Figs. 8-21, 8-22 and 8-23).

A similar installation is at the Georgia Power Headquarters in Atlanta. There, the solar system was designed to displace 15 percent of the total energy requirements of the office building. Up to 30 percent of the energy required

Fig. 8-22. Honeywell Plaza, detail of parabolic collectors.

Fig. 8-23. Hammel Green Abrahamson, Inc.: Honeywell Plaza, Minneapolis, 1978. Collectors shade the cars while capturing the sun.

Fig. 8-24. Heery and Heery, Architects and Engineers: Georgia Power Company Headquarters, Atlanta.

Fig. 8-25. Crosbyton Solar Power Project, featuring ten 200-foot diameter solar "bowls," expected to produce 5 megawatts of power.

Selected Bibliography

Anderson, Bruce with Riordan, Michael. *The Solar Home Book.* New Hampshire: Cheshire Books, 1976.

Blake, Peter. *Form Follows Fiasco.* Boston: Little, Brown and Company, 1977.

Blondel, Jacques Francois. *De La Distribution Des Maisons De Plaisance.* Vol. II. Paris: Charles-Antoine Jombert, 1738.

Collins, Belinda L. *Windows and People: A Literature Survey.* Washington, D.C.: U.S. Government Printing Office, 1975.

Drexler, Arthur. *Transformations in Modern Architecture.* New York: The Museum of Modern Art, 1979.

Fine, Oronce. *Opere di Orontio Fineo del Delfinato.* Venice: F. Franceschi, 1587.

Fisher, Rick and Yanda, Bill. *Solar Greenhouse.* New Mexico: John Muir Publications, 1976.

Giedion, Siegfried. *Space, Time and Architecture,* Fifth Ed. Cambridge: Harvard University Press, 1967.

Gluck, Irvin D. *It's All Done With Mirrors.* New York: Doubleday and Company, Inc., 1968.

Greif, Martin. *Depression Modern—The Thirties Style in America.* New York: Universe Books, 1975.

Hitchcock, Henry Russell and Johnson, Philip. *The International Style.* New York: W.W. Norton and Co., Inc., 1966.

Hopkins, Albert A. *Magic.* New York: Munn and Co., 1901.

Jencks, Charles. *The Language of Post-Modern Architecture.* New York: Rizzoli International Publications, Inc., 1977.

Lambeth, James and Delap, John Darwin (ed.). *Solar Designing.* U.S.A.: James Lambeth, 1977.

Lasch, Christopher. *The Culture of Narcissism.* New York: W.W. Norton and Co. Inc., 1979.

Lesieutre, Alan. *The Spirit and Splendor of Art Deco.* New York: Paddington Press, 1974.

McGrath, Raymond and Frost, A.C. *Glass in Architecture and Decoration.* London: Architectural Press, 1961.

Menten, Theodore. *The Art Deco Style.* New York: Dover Publications, Inc., 1972.

Montgomery, Richard H., with Budnick, Jim. *The Solar Decision Book.* Midland, Michigan: Dow Corning Corporation, 1978.

Moore, Charles, Allen, Gerald and Lyndon, Donlyn. *The Place of Houses.* New York: Holt, Rinehart and Winston, 1974.

Newman, Jay Hartley and Newman, Lee Scott. *The Mirror Book.* New York: Crown Publishers, Inc., 1978.

Pope, Charles Henry. *Solar Heat, its Practical Applications.* Boston: C.H. Pope, 1903.

Repton, Humphrey assisted by son Repton, J.A. *Fragments on the Theory and Practice of Landscape Gardening.* London: T. Bensley and Son, 1816.

Robinson, Cervin and Bletter, Rosemarie Haag. *Skyscraper Style—Art Deco New York.* New York: Oxford University Press, 1975.

Saarinen, Aline. *Eero Saarinen on his Work.* New Haven: Yale University Press, 1962.

Smith, C. Ray. *Supermannerism.* New York: E.P. Dutton, 1977.

Stobaugh, Robert and Yergin, Daniel, (eds.). *Energy Future.* New York: Random House, 1979.

Venturi, Robert. *Complexity and Contradiction in Architecture.* New York: The Museum of Modern Art, 1977.

Voltaire; Trevor-Roper, Hugh R. (ed.). *The Age of Louis XIV.* New York: Washington Square Press, Inc., 1963.

Wharton, Edith and Codman, Ogden, Jr. *The Decoration of Houses.* New York: W.W. Norton and Co., Inc., 1978.

PERIODICALS

Extensive research was done in American periodicals, starting at the end of the nineteenth century and proceeding to the present. Articles cited here seem particularly relevant, or were quoted at length in text.

Birkerts, Gunnar. "Defining a Design Method," *Architectural Record* (February, 1977).

Bottomly, Wm. Lawrence. "Mirrors in Interior Architecture." *Architectural Forum* (October, 1932).

Dahl, J.O. "Repeal Offers New Opportunities." *Architectural Record* (December, 1933).

Guntner, C. "Utilization of Solar Heat by a Plane Mirror Reflector." *Scientific American* (May 26, 1906).

Hitchcock, H.R. "Some American Interiors in the Modern Style." *Architectural Record* (July, 1928).

Huxtable, Ada Louise. "Skyscrapers, a 'New' Esthetic and Recycling." *The New York Times* (December 26, 1976).

Johnson, Philip. "A There There," *Architectural Forum* (November, 1973).

Lessard, Susan. "A Reporter at Large—The Towers of Light," *The New Yorker* (July 10, 1978).

"New Products—Mirror, Mirror on the Wall," *Business Week* (May 7, 1966).

"Palace of Mirages," *Scientific American* (May 1909).

Pelli, Cesar. "Transparency—Physical and Perceptual," *A + U, Architecture and Urbanism* (November, 1976).

Wright, Frank Lloyd. "The Meaning of Materials—Glass." *Architectural Record* (July, 1928).

Index